Rural Community Water Supply

Praise for this book

'Richard Carter has written a most valuable and detailed account of rural water supply which covers a wide range of subjects and contains a huge list of valuable references. It should find its way and be read by a large number of people who are studying, working or linked to this important discipline.'

Peter Morgan, Consultant and Stockholm Water Prize Winner

'This book is simply excellent – for those starting out, for those wanting to know more, and for those who have been in the sector for years. I appreciate it for its pragmatism, for being comprehensive, providing incredible detail as well as history, and looking into the future. To quote the author, "There is still a long way to go to achieve basic, on premises and safely managed services". I believe that this book will become a classic, a resource and reference for those striving to improve rural community water supplies – for everyone, everywhere.'

Dr Kerstin Danert, Ask for Water GmbH, Switzerland

'This book is a significant resource for anyone working in the rural water sector in the run up to the SDG target date of 2030. It is truly impressive in scope and helps to debunk some of the myths around rural water provision, as well as reclaiming approaches that some see as being written off too easily in the past. The historic perspective of the book reflects Richard's long involvement in – and passion for – improving water services for rural people around the world and serves to remind us that some lessons are timeless. Finally, the author's human spirit shines through as he consistently puts people and power dynamics at the heart of proposed solutions, as much as engineering and money.'

Harold Lockwood, Director, Aguaconsult UK

'Richard Carter has created an impressive resource for all those committed to improving the lives of rural dwellers. He's brought together an extraordinary amount of information, brought it up to date, and presented it in a well-organized, actionable form. Everything is clearly explained, and new evidence is integrated with well-established science. The emphasis on practical steps to achieve progressive improvement of rural water supply systems is valuable for both the practitioner and the policymaker.'

Clarissa Brocklehurst, Gillings School of Global Public Health; Water Institute at The University of North Carolina at Chapel Hill

'This book is a "must read" for those interested in rural water supply. It is not only a much needed update on technologies since "The handpump option" (World Bank, 1987) but also gives a wealth of experiences in non-technical aspects to make rural water supply sustainable. Improving access in rural areas is not only about reaching SDG 6.1 but it also contributes to water-related SDGs concerning income, food, and employment.'

Henk Holtslag, WASH specialist

'An authoritative and multidisciplinary review of the historical performance and future prospects to address the enduring global challenge of delivering drinking water to rural people.'

Professor Rob Hope, University of Oxford

'This book is a mine and a wealth of resources for professionals, students, and adherents of rural water supply. Against a historical context, Prof. Carter outlines the practical steps needed to improve rural water supply for rural people in low and middle-income countries. In the words of Nelson Mandela, "a nation should not be judged by how it treats its highest citizens but its lowest ones". Carter expounds this by bringing out the latest, proven, and best practices available in rural water supply from the perspective of the poorest communities in rural areas.'

Javan Nkhosi, water engineer, consultant, and author, Zambia

Rural Community Water Supply
Sustainable services for all

Richard C. Carter

Practical Action Publishing Ltd
27a Albert Street, Rugby,
CV21 2SG, UK
www.practicalactionpublishing.com

© Richard C. Carter, 2021

The moral right of the author to be identified as author of this work and the contributors to be identified as contributors of this work have been asserted under sections 77 and 78 of the Copyright Designs and Patents Act 1988.

This open access book is distributed under a Creative Commons Attribution Non-commercial No-derivatives CC BY-NC-ND licence. This allows the reader to copy and redistribute the material; but appropriate credit must be given, the material must not be used for commercial purposes, and if the material is transformed or built upon the modified material may not be distributed. For further information see https:// creativecommons.org/licenses/by-nc-nd/4.0/legalcode

Product or corporate names may be trademarks or registered trademarks, and are used only for identification and explanation without intent to infringe.

A catalogue record for this book is available from the British Library.
A catalogue record for this book has been requested from the Library of Congress.

ISBN 978-1-78853-165-8 Paperback
ISBN 978-1-78853-166-5 Hardback
ISBN 978-1-78853-168-9 Ebook

Citation: Carter, Richard C. (2021) *Rural Community Water Supply: Sustainable services for all*, Rugby, UK: Practical Action Publishing <http://dx.doi.org/10.3362/9781788531689>.

Since 1974, Practical Action Publishing has published and disseminated books and information in support of international development work throughout the world. Practical Action Publishing is a trading name of Practical Action Publishing Ltd (Company Reg. No. 1159018), the wholly owned publishing company of Practical Action. Practical Action Publishing trades only in support of its parent charity objectives and any profits are covenanted back to Practical Action (Charity Reg. No. 247257, Group VAT Registration No. 880 9924 76).

The views and opinions in this publication are those of the author and do not represent those of Practical Action Publishing Ltd or its parent charity Practical Action. Reasonable efforts have been made to publish reliable data and information, but the authors and publisher cannot assume responsibility for the validity of all materials or for the consequences of their use.

Cover photos: (top) installing borehole casing and screen into a newly drilled borehole in Lira District, northern Uganda; (bottom) fetching water from an Afridev handpump in Blantyre District, Malawi.
Credit: Richard C. Carter
Cover design: Brian Melville
Typeset by vPrompt eServices, India
Printed in the United Kingdom

Contents

Boxes, figures, and tables	vi
Acronyms	xi
Acknowledgements	xiii
Preface	xv
About the author	xvi
Who this book is for	xvii
My intention in this book	xviii
Covid-19	xix
Black lives matter	xxi
1. Sustainable rural water services for all	1
2. Water quantity, quality, and health	13
3. Groundwater resources	29
4. Water supply boreholes	49
5. Water lifting from wells and boreholes: handpumps	67
6. Water supply infrastructure: beyond handpumps	87
7. From getting it going to keeping it flowing: management of rural water services	105
8. Finance: the fuel for sustainable rural water services	121
9. Rural water users and community water supply programmes	135
10. Water for all: why is it such a struggle, and what can be done?	151
11. What's changing in rural water supply?	167
12. Imagine another world	185
Endnote: National WASH systems sit within a global system of injustice	199
Annex: Some notes on definitions and statistics	201
References	207
Index	239

Boxes, figures, and tables

Boxes

4.1	Nine principles for cost-effective boreholes	51
4.2	Ten key factors in well or borehole site selection	52
4.3	Important aquifer properties	56
4.4	Estimating drawdown and pumping water level during the design process	56
4.5	Test pumping	63
4.6	Water quality sampling	64
4.7	The supervisor's responsibilities	66
6.1	Head and the hydraulic grade line in piped systems	95
7.1	Extract from the 'Mar del Plata Action Plan' in relation to community water supply	109
7.2	The Paris Principles on aid effectiveness	111
7.3	Eight attributes required of rural community water management arrangements	112
7.4	An example of well-performing community management: Kigezi Diocese, Uganda	117
7.5	Community-managed handpumps in Malawi: the Madzi Alipo programme	119
9.1	What it means to be multi-dimensionally poor	139
9.2	Some key principles regarding community engagement	146
9.3	Typical features of rural community water projects	147
10.1	The essence of the human right to water	162
10.2	Essential elements of democracy	163
11.1	Some weaknesses of community management	180
11.2	Systems – complicated and complex	184

BOXES, FIGURES, AND TABLES   vii

Figures

1.1	Rural population projections, 1980–2050, for major world regions	3
1.2	Simplified schematic showing spectrum of primary dwellings from rural to urban	5
1.3	The scope of this book	10
2.1	Trends in water services, 2000–17, Central and Southern Asia and sub-Saharan Africa, % of rural populations	14
2.2	A simplified form of the F-diagram, incorporating animal excreta	18
2.3	Health aspects of inorganic chemical constituents of groundwater	21
3.1	A simplified version of the water cycle	31
3.2	Examples of application of water balance principles	32
3.3	Rainfall variability from year to year, 1961–97	34
3.4	Potential and actual evapotranspiration rates	35
3.5	Indirect recharge from a watercourse to an adjacent aquifer	36
3.6	Estimated cumulative groundwater depletion by region since 1900, km^3	37
3.7	Key features of unconfined and confined aquifers	39
3.8	Hydraulic conductivity ranges of selected rock types, meters per day	40
3.9	The relationship between hydraulic conductivity and water content	42
3.10	The cone of depression or drawdown curve around a well or borehole	43
3.11	Practical limitation on groundwater abstraction rate in an unconfined aquifer	43
3.12	Population projections (billion people), 2000–50, by region and within sub-Saharan Africa	46
4.1	Main features of a borehole drilled in an unconsolidated formation	54
4.2	Flow chart summarizing the borehole design process	55
4.3	Main water borehole drilling methods	60
4.4	Borehole straightness (alignment) and verticality (plumbness)	62
4.5	The desired outcome of well development	63

5.1	Water pump components, conceptually and practically	68
5.2	The Bucket Pump and the Rope Pump	71
5.3	The Canzee pump and the EMAS pump	72
5.4	Three 'revolutionary' deep well handpumps	75
5.5	The Volanta and the Blue Pump	76
5.6	The Vergnet and Mono pumps	78
5.7	Corrosion of India Mark II galvanized rising main in northern Uganda	83
6.1	Simple branched gravity flow schemes	89
6.2	Mechanical pumping infrastructure	91
6.3	Submersible electric borehole pump (schematic)	92
6.4	Submersible pump characteristic curves	93
6.5	The total dynamic head (a+b+c) for a submersible pump in a borehole or well	94
6.6	The Impact Pump	99
6.7	Head and hydraulic grade line in a simple piped system	102
6.8	Public standposts and water kiosks	103
7.1	WASH expenditure as a percentage of GDP and per capita for 35 countries	110
7.2	Necessary actions before and in response to handpump breakdown	118
8.1	Cost proportions for different cost components	125
8.2	WASHCost average costs compared: handpumps vs piped systems	126
8.3	Rural water annual costs per person (US$ 2015) by service and cost component	129
8.4	Benefit/cost ratios of investments to achieve universal access to improved drinking water by 2015	131
9.1	Rural population growth and projections 2000–50, billions	142
10.1	Local, national, and international checks and balances	158
10.2	Opposing negative individual attributes	159
10.3	The logic of institutional reform	165

11.1	Service levels and change, 2000–17, rural water supply, global	169
11.2	Bottom 49 countries ranked by 'at least basic' service in 2017	172
11.3	Gross national income per capita vs rural water 'at least basic' access (%), 2017	173
12.1	Schematic representation of a phased approach to rural water services, Uganda	193

Tables

1.1	JMP definitions of water services in the SDG era (2016–30)	4
1.2	Global and regional rural drinking water coverage, 2017, % and millions served	5
2.1	Rural domestic water requirements	15
2.2	JMP service ladders for WASH in schools	25
2.3	JMP service ladders for WASH in health care facilities	27
3.1	Aquifer potentiality as a function of aquifer transmissivity	43
3.2	Well or borehole yield and use	45
4.1	Design discharge, pump outer diameter (o.d.), and casing internal diameter (i.d.)	57
5.1	Main types of water pump used in rural water supply	70
5.2	Country statistics for functionality	84
6.1	Water delivery system components compared	96
6.2	Capex, opex, and capmanex costs of solar pumping schemes of different sizes	96
8.1	Life-cycle costing: main cost components	124
8.2	Summary of unit cost assumptions made by Hutton and Varughese (2016)	126
8.3	Country-averaged per-person costs in Hutton and Varughese (2016) model	127
9.1	Global numbers of people lacking access to improved and safely managed domestic water services, 2017 data	138
9.2	Numbers of people in low-income countries experiencing multi-dimensional poverty	139

9.3	Numbers of people in lower-middle income countries experiencing multi-dimensional poverty	140
9.4	Achieving equality in rural water programmes	148
11.1	Service levels and change, 2000–17, rural water supply (global)	168
11.2	Bottom 49 countries ranked by 'at least basic' (sum of basic and on premises) in 2017 (percentage access by rural population)	170
11.3	Recent technology developments in rural water supply	175
11.4	Water utility indicators	178
11.5	Suggested rural water supply performance indicators	179
12.1	Rural water service level access, 2000 and 2017, Uganda	193
A.1	World Bank country classifications: low- and middle-income countries	202
A.2	The SDG geographical regions	204

Acronyms

ACF	Action contre la faim
AMCOW	African Ministers' Council on Water
ASTM	American Society for Testing and Materials
BGS	British Geological Survey
Capex	Capital expenditure
Capmanex	Capital maintenance expenditure
CDC	Centers for Disease Control and Prevention
CESR	Center for Economic and Social Rights
CV	Coefficient of variation
DFID	Department for International Development
EED	Environmental enteric dysfunction
ET	Evapotranspiration
ExpDS	Expenditure on direct support
ExpIDS	Expenditure on indirect support
FAO	Food and Agriculture Organization of the United Nations
GFS	Gravity flow schemes
GLAAS	Global Analysis and Assessment of Sanitation and Drinking Water
GNI	Gross national income
HDPE	High-density polyethylene
HPB	Boreholes equipped with handpumps
HWTS	Household Water Treatment and Safe Storage
IAH	International Association of Hydrogeologists
IBNET	International Benchmarking Network
ICESCR	International Covenant on Economic, Social and Cultural Rights
IDPG	International Development Partners Group, Nepal
INGO	International non-governmental organization
IPCC	Intergovernmental Panel on Climate Change
JMP	Joint Monitoring Programme
KDWSP	Kigezi Diocese Water and Sanitation Programme
LLC	Life-cycle costing
MDG	Millennium Development Goals
MDPI	Multi-dimensional poverty index
MOH	Ministry of Health
MOWS	Ministry of Works and Supplies
NGO	Non-government organization
NICC	Netherlands International Cooperation Collection
NTD	Neglected tropical diseases
OD	Open defecation
ODA	Official Development Assistance
ODK	Open Data Kit
OECD	Organisation for Economic Co-operation and Development
OHCHR	Office of the United Nations High Commissioner for Human Rights
Opex	Operational expenditure
OPHI	Oxford Poverty and Human Development Institute

O&M	Operation and maintenance
PBR	Payment by Results
RWSN	Rural Water Supply Network
SDG	Sustainable Development Goals
SIWI	Stockholm International Water Institute
UN	United Nations
UNDESA	United Nations Departments of Economic and Social Affairs
UNEP	United Nations Environment Programme
UNICEF	United Nations International Children's Emergency Fund
UPGro	Unlocking the Potential of Groundwater for the Poor
USAID	United States Agency for International Development
VLOM	Village-level operation and maintenance
VLOM(M)	Village-level operations and (management of) maintenance
WASH	Water, sanitation, and hygiene
WEDC	Water Engineering and Development Centre
WHO	World Health Organization
WPDx	Water Point Data Exchange
WSP	Water and Sanitation Programme

Acknowledgements

This book is in large part a reflection on four-and-a-half decades of working with communities, national and local governments, non-governmental organizations, donors, international agencies, private sector organizations, students, colleagues, and a wide range of development professionals. All have contributed to my own learning in a multitude of ways – most of which I can no longer individually identify and attribute. I am so grateful to have lived this learning journey.

I was based for the single longest period of my working life at Silsoe College, Cranfield University. While there I established a Master's programme in Rural Water Supply. Despite title and curriculum changes it continues more than 30 years later. In 2016 it received a Queen's Award for Higher Education. I cannot say how grateful I am to the hundreds of MSc students and dozens of PhD students with whom teaching, discovery, and learning were mutual.

Many professional colleagues have helped me along the way, at Silsoe, at WaterAid, through the Rural Water Supply Network, in Ethiopia, Madagascar, Malawi, Nigeria, Uganda, and numerous other African and Asian countries.

I have been involved with the journal *Waterlines* and its publishers, Practical Action Publishing (formerly ITDG,) since its first issue in 1982. I have served as Editor since 2008. It has been a pleasure to work with and learn from my colleagues there.

The manuscript for this book was professionally content and copy-edited by Andrea Scheibler and peer-reviewed by Vincent Casey, Peter Harvey, Henk Holtslag, Sampath Kumar, and Peter Morgan prior to it reaching the publishers. I am extremely grateful to those individuals for the work done to improve the accuracy, flow, and content of the narrative. I am very grateful, too, to those individuals at Practical Action Publishing (Clare Tawney, Chloe Callan-Foster, Kelly Somers, Andrea Johnson, Jenny Peebles, Katarzyna Markowska, and Rosanna Denning, and proofreader Louise Medland) who have worked hard to bring this book to fruition.

The publication of this book in an open access form has been made possible by generous support from (a) the British Geological Survey and the Hidden Crisis research project (grant number NE/M008606/1) within the UPGro programme, co-funded by the Natural Environment Research Council (NERC), the Foreign, Commonwealth and Development Office (FCDO), and the Economic and Social Research Council (ESRC); and (b) the Vitol Foundation. I am very grateful for this financial support.

The list of individuals who have, knowingly or not, helped to mould my professional thinking in various ways over the last 45 years is long, and despite the risk of omitting someone, I wish to particularly acknowledge and thank the following: Ian Acworth, Len Abrams, Bill Adams, Dotun Adekile,

A.B. Alhassan, Alhaji Goni Alkali (deceased), Ian Alsop, Adisse Amado, George Bagamuhunda, Peter Ball, Samson Bekele, Kenneth Bekunda, Jane Bevan, Manfred Bojang, Louis Boorstin, Clarissa Brocklehurst, Willy Burgess, Vincent Byarugaba, Mike Byers, Reuben Byomuhangi, Sandy Cairncross, George Cansdale (deceased), Richard Cansdale, Roger Calow, Ken Caplan, Mike Carr (deceased), Vincent Casey, Sue Cavill, Robert Chambers, Clive Chapman (deceased), Wiktor Chichlowski, John Chilton, Frances Cleaver, Jeremy Colin, Oliver Cumming, Clare Cummings, Val Curtis (deceased), Michael Assefa Dadi, Robin D'Alessandro, Lucien Damiba, Kerstin Danert, Ian Davis, St John (Singe) Day, Desta Demessie, Julian Doczi, Jane Dottridge, Peter Dumble, Sunil Dutta, Mike Edmunds (deceased), Victor Eilers, Jeroen Ensink (deceased), Barbara Evans, Richard Feachem, Fabrizio Fellini, Dick Fenner, Gareth and Ursula Firth, Tim Forster, Stephen Foster, Tim Foster, Richard Franceys, Sean Furey, Pam Furniss, Régis Garandeau, Om Prasad Gautam, Silvia Gaya, Louisa Gosling, Frank Greaves, Mohammed Hamza, Mike Hann (deceased), Erik Harvey, Peter Harvey, Muhammad Hassan, Robin Temple Hazell (deceased), Tim Hess, Rob Hope, Peter Howsam, Jim Houston, Paul Hunter, Paul Hutchings, Nigel Janes, John Jewsbury, Guy Jobbins, Stephen Lindley-Jones, Glyn Jones, Steve Jones, Jeremy and Margareta Jose, Robert Kampala, Melvyn Kay, Aaron Kabirizi, Gilbert Kimanzi, Bruce Lankford, Dan Lapworth, Adrian Laycock, Harold Lockwood, Eva Ludi, Alan MacDonald, Brian Mathew, Duncan McNicholl, Bruce Mead, Andy Meakins (deceased), Mogus Mehari, Mario Milanesi, Bruce Misstear, Dai Morgan, Peter Morgan, Patrick Moriarty, Mike Mortimore (deceased), Masauko Mthunzi, Francis Musinguzi, Isaac Mutenyo, Mike Mutter, Jon Naugle, Ian Neal (deceased), Tommy Ngai, Muthi Nhlema, Henry Northover, Grace Oluwasanya, Christie Peacock, Tony Preston, Peter Ravenscroft, Colin Richardson, Peter Robbins, Jan Willem Rosenboom, Ian Ross, Ray Rowles (deceased), Ken Rushton, Ronnie Rwamwanja, Peter Ryan, Janusz Rydzewski (deceased), Ton Schouten (deceased), Vitor Serrano, Tom Slaymaker, Ron Sloots, Jen Smith, Jamil Ssebalu, Peter Stern (deceased), David Sutherland, Sally Sutton, Clare Tawney, Richard Taylor, John Thompson, Lucrezia Tincani, Callist Tindimugaya, Jérémie Toubkiss, Andy Trevett, Mark Trigg, Sean Tyrrel, Harry Underhill, Paul Venton, Anthony and Juliet Waterkeyn, Keith Weatherhead, James Webster, Luke Whaley, Tal Woolsey, Ed Wright (deceased), Jeff Yoder, Paul Younger (deceased), and Mark Zeitoun.

Finally my deep gratitude to my wife Maryla and my family for putting up with me, my travels, and my obsession with my life's work over many decades, and supporting me in so many ways.

Preface

The Sustainable Development Goals (SDGs) include the target of 'safely managed' water services for all by 2030. Safely managed water refers to a service that is provided on the premises (piped or otherwise), available on-demand at least 12 hours out of every 24, and of a quality that matches national or World Health Organization standards.

Many governments, donors, private sector, and non-governmental organizations are working to achieve this laudable target, but for many rural households and communities it will not be met – either within the final decade of the SDG period, or even shortly afterwards. Because of the inherent management and financing obstacles, the target will remain a pipe dream.

Given this reality, what should be done? The lamentable situation of those who at present rely on contaminated surface water, unprotected groundwater, or protected-but-still-distant groundwater sources must be relieved.

While not disagreeing with the SDG ambition, I argue in this book for a realistic and pluralistic approach to rural water services, enabling people to move up the ladder of improving service levels. To maintain this momentum I place a heavy emphasis on solving the sustainability challenge. In other words, once communities enjoy a higher level of service, they should either continue to do so, or move on up, but certainly not regress.

Addressing the rural water service challenge of the 2020s and beyond requires a greater depth of understanding from across the natural sciences and engineering, social science, political, and economics disciplines. It requires solutions that fit the contexts of countries and sub-national regions. It needs determination, investment, and new ways of managing and financing services. New technology has its part to play, too.

Rural water service provision is a fast-changing, dynamic, exciting, and rewarding field of human endeavour. It is my hope that this book will help to spur progress towards realistic and increasingly effective outcomes for all rural people, especially for those who remain at risk of being left behind in national and international attempts to relieve water poverty and provide domestic water security.

Richard C. Carter

About the author

My undergraduate studies were in the natural sciences, with a specialism in geology. It was this specialization which initiated my later interest in groundwater. On completing my first degree, which had been strong on theory but less so on applied aspects, I undertook a taught Masters programme in Irrigation and Water Resources Engineering. This exposed me to engineering aspects of water development and led me into theoretical studies of groundwater. My working life, which has been a mix of consultancy, research, and teaching, has taken me to many countries in Africa and Asia. My interests have always been focused on water, especially in rural areas and especially among the most disadvantaged communities in low-income countries. Water is so fundamental to household life and food security, and so closely linked to poverty, and it is these links that have motivated me over the last 45 years of my professional life. To the extent that I have any understanding of rural sociology, culture, politics, economics, and institutions, it is this intensely human aspect of rural water supply that has helped to educate me.

Who this book is for

This book is intended to be read by, or to influence, two groups of people. The first group includes national and local government personnel, non-governmental organization programme staff and volunteers, students, and others who are (or intend to be) directly involved in the implementation of rural water programmes. For you, I hope this book will provide a helpful overview of both the state of knowledge and the 'state of the art' of rural water services as we embark on the final decade of the Sustainable Development Goals. Whether you read the book from cover to cover, or simply refer to it from time to time, my wish is that the book will help you in the ways you plan, execute, and evaluate your work.

The second group consists of those who are a little more distant from the day-to-day work of rural water service provision, and who may not require highly specialized knowledge. I am thinking here of those who make decisions about the funding of development work, and especially those who help to determine how budgets in the water, sanitation, and hygiene sector are allocated and spent. I would be delighted if this book sits on the desks (or in the computers) of such individuals and organizations, and that you dip in from time to time.

My intention in this book

I have deliberately kept the scope of the book broad; this has inevitably limited how much detail I could go into on any particular aspect. I have tried to explain ideas, concepts, and sector thinking in plain English, and so make all the subject matter of the book accessible. Whether your own background is in the natural or social sciences, in engineering or technology, I hope there is something in this book with which you can connect – or perhaps that the 'bigger picture' that I have tried to paint may be the thing that makes this book worthwhile.

Whatever the case, and whoever you are, it is my hope that this book can contribute in a small way to improving the lives of rural people in low- and middle-income countries.

Covid-19

During the writing of this book, a new disease-causing virus, SARS-CoV-2, and the disease it causes, Covid-19, entered the experience of every nation and person on the planet. The first case is thought to have occurred in mid-November 2019 in Wuhan, China, although it may have existed for longer. The disease rapidly spread around the world, and on 11 March 2020 the World Health Organization declared it a global pandemic. As a consequence, many countries introduced lockdowns to limit transmission; at the same time attention turned to what other measures could be taken to manage morbidity and mortality.

At the time of writing, this respiratory disease has no cure. Drugs have been identified that can relieve the symptoms in those affected most seriously; and newly developed vaccines are being deployed. Mortality rates increase with age (only rising above 1 per cent in over-50s and exceeding 10 per cent in those in their late 70s and those over-80; see Ruan, 2020), especially in patients with pre-existing conditions such as cardiovascular disease, chronic kidney disease, diabetes, or chronic respiratory disease; men are more susceptible than women (Clark et al., 2020).

One of the most effective ways in which individuals can protect themselves from infection is through frequent and thorough handwashing with soap. However, if people lack access to sufficient water for that purpose, or if they are too poor to buy soap, this simple measure is beyond their means.

The communities and households that are the focus of this book are precisely those who lack reliable access to sufficient quantities of acceptably safe domestic water. They live in low-income or lower-middle income countries. They are often among the poorest individuals in those countries. Many of them have very limited disposable cash, as the greater part of their incomes arise from subsistence farming, with little in the way of surpluses for market.

Many organizations that are actively addressing the disease and its impacts, and many individuals involved in research on Covid-19, have drawn attention to the disproportionate impacts of the disease and of lockdown on poor people in low-income countries (Broadbent et al., 2020; Kelley et al., 2020; UNICEF, 2020; World Bank, 2020a).

The outbreak of Covid-19 was not predicted in specific terms, but the likelihood of a global pandemic has been known about for many years, and the possibility of moving 'from anecdotal through analytical to potentially predictive' mode was highlighted as imminent nearly 10 years ago (Morse et al., 2012: 1963).

Having sufficient and reliable water, which is of good enough quality, close by, manageable, and affordable, is one of the key ways of mitigating the impacts of many infectious agents that cause diarrhoeal or respiratory disease. Covid-19 is simply the most recent, and spectacular, addition to a long list of human pathogens. However, its impact on national gross domestic product, food security, and the poverty of those on low incomes in low- and middle-income countries will persist for many years, setting back fragile progress towards the achievement of the Sustainable Development Goals.

Black lives matter

During the writing of this book, renewed support for the Black Lives Matter movement was catalysed by the police murder of George Floyd in Minneapolis, USA, in May of 2020. In solidarity with the numerous other Black victims of police brutality, there followed a global wave of protests against the institutionalized racism inherent in many state apparatuses.

This systemic racism has been likened to the Covid-19 pandemic: 'We are living in a racism pandemic, which is taking a heavy psychological toll on our African American citizens' (Shullman, 2020). The actual link is even stronger than a metaphorical one, as evidence from Public Health England (2020) is clear that the Covid-19 outbreak is disproportionately affecting Black, Asian, and Minority Ethnic people (often referred to in the UK as BAME). The pandemic, and other health and social welfare indicators, starkly reveals the continued disadvantages that black and brown citizens in the USA, UK, and many other white-majority countries must contend with on a daily basis.

The reason for raising this issue in the preliminaries to this book is simple and uncomfortable. The majority of people around the world who lack safe and sustainable drinking water – located primarily in Africa and Asia, and among the indigenous communities of Australia, Canada, and the United States – share two things in common. First, the colour of their skin – most are non-white; and second, their historical subjugation by European colonial and neo-colonial systems. The most extreme expressions of colonial abuse of power included slavery, segregation, and genocide; but the subtler exclusion from peoples' opportunities to fulfil their potential was arguably as violent and corrosive – particularly in their long-term impacts, as the consequential poverty experienced by colonized populations persists to this day.

One way in which racial disadvantage is embedded in systems of research, thought-leadership, funding, and goal setting in the water sector is through the continued dominance of white (mostly male) faces in positions of influence and power. This was recently highlighted by the use of the term 'helicopter (or parachute, or neo-colonial) research' – in which '[mainly white] researchers from wealthier countries fly to a developing country…and publish the research with little involvement from local scientists' (Minasny et al., 2020; see also Bates et al., 2020). Similar arguments were being made during the drafting of this book for the decolonization of the WASH sector (Luseka, 2020).

There is a special and urgent imperative to see real progress toward equality of opportunity and well-being in this generation. This means understanding the systemic factors that cause inequity and inequality, and working intelligently and tirelessly to put them right. Only by doing so can the SDG mantra 'leave no-one behind' be fulfilled in a meaningful way.

CHAPTER 1
Sustainable rural water services for all

Abstract: *Approximately two-thirds of a billion rural people, mostly in sub-Saharan Africa and Asia, collect their domestic water from unprotected sources or from engineered but nonetheless distant water points. Another three-quarters of a billion rural people enjoy a 'basic' water supply from engineered water points outside the home but within a 30-minute round trip; however, the reliability and safety of the service is often inadequate. The reasons for such a lamentable situation are numerous and complex; their solutions lie in improvements to the ways rural water services are governed by nation states, the financial and management arrangements by which services are kept working, the behaviours of water users, and the technologies that supply and monitor water. The purpose of this book is to set out – for governments, donors, and non-governmental organizations – the nature of the water problems faced by disadvantaged rural communities and, more importantly, to point to solutions and strategies that would ameliorate access to and management of water resources during and beyond the final decade of the UN's Sustainable Development Goals.*

Keywords: rural water supply, domestic water supply, drinking water, community water supply, community-managed water, groundwater, water points, Sustainable Development Goals

'Some for all rather than more for some'
—United Nations, 1990

Progress toward sustainable rural domestic water services for all

The first concerted global effort to serve the entire world's population with adequate domestic water services was the UN International Drinking Water Supply and Sanitation Decade of the 1980s. The backdrop to this initiative was the awareness in the late 1970s of high mortality resulting from lack of safe water and adequate sanitation – a figure of more than 30,000 deaths per day was cited by Arlosoroff et al. (1987); and the estimate that more than 70 per cent of the world's rural population lacked access to a safe and adequate water supply.

In the period since the Water Decade, much progress has been made. Death and disease due to poor water and sanitation have reduced sharply. Diarrhoea mortality in under-fives has decreased by nearly 70 per cent between 1990 and 2017 (Global Burden of Disease Collaborators, 2019). According to the UNICEF and World Health Organization (WHO) Joint Monitoring Programme (JMP), the proportion of rural people globally who still lack an improved drinking water service is down from more than 70 per cent to 15 per cent

(JMP, 2019). The Millennium Development Goal (MDG) target was met in 2012. All of this is good news.

And yet major challenges persist. Still almost 1,500 children under five years of age are estimated to die every day from diarrhoeal disease, much of which is related to poor water and sanitation. Although the MDG target referred to safe water, the proxy indicator it used for this (access to an 'improved' water source) is widely recognised to be imperfect. Other contributors to these disease statistics include poor hand and food hygiene, and environmental contamination from human and animal pathogens. Far greater recognition is given nowadays to the importance of sanitation and hygiene, and to the changes in behaviour and practice that are necessary to bring about significant health impacts, especially in regard to stunting (impaired growth resulting from inadequate nutrition and recurrent infections). However, achieving those desired health impacts in practice still often proves elusive (Schmidt, 2014; Luby et al., 2018; Null et al., 2018; Cumming et al., 2019; Humphrey et al., 2019).

Global rural population has increased by 26 per cent between 1980 and 2020, although as country populations increasingly urbanize these numbers are expected to fall. However, global totals hide important regional differences. While rural population numbers in Asia, Europe, Latin America and the Caribbean, Northern America, and Oceania have all peaked and are now falling, Africa's rural population (and especially that of sub-Saharan Africa) is projected to continue rising to 2050 and beyond (Figure 1.1).

The MDG target to 'reduce by half (by 2015, relative to 1990) the proportion of people lacking sustainable access to safe drinking water' – would not bring services to all, and both the safety (water quality) and sustainability dimensions of the services delivered have been widely criticized for their shortcomings (Bain et al., 2014; Foster et al., 2019). Many rural people have made or received a first-time improvement to their drinking water supply, but standards of construction and water quality and shortcomings in the level and reliability of services still leave much to be desired.

Community management of rural drinking water services

During and after the Water Decade, the idea of **community participation** evolved into the principle of **community management** of rural water services. Although it was assumed that capital investments in new facilities would be made by governments and their development partners (i.e. donors and NGOs), it gradually became accepted – though not always clearly explained to water users – that new infrastructure would be 'handed over' to communities, who would be deemed responsible for the on-going management and post-construction financing of the services provided by drinking water infrastructure (the protected springs, wells, and boreholes, and piped systems and public tapstands that deliver water). Communities were expected to take responsibility for repairs to masonry, handpumps, pipelines, and taps, arranging the necessary services, supplies, and finance to do so. The present-day paradigm

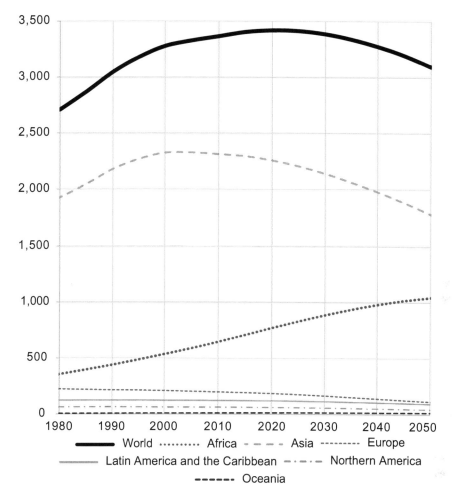

Figure 1.1 Rural population projections, 1980–2050, for major world regions
Source: UN DESA, 2019c

of community management had its origins prior to the 1980s, but it was furthered significantly during the Water Decade and in high-profile international conferences following that period. It remains the dominant – but increasingly challenged – model for rural drinking water services today (Schouten and Moriarty, 2003; Whaley et al., 2019; Hope et al., 2020).

Water service levels

The ideal of all people having on-demand access to a safe and dependable supply of piped water in the home is well recognized. It is encapsulated in the Sustainable Development Goal (SDG) target 6.1, the goal of which is for all people to enjoy 'safely managed' drinking water services by 2030 (SDG Tracker, n.d.).

Table 1.1 JMP definitions of water services in the SDG era (2016–30)

Service level	Description	Comment
Safely managed	Drinking water from an improved water source that is located on premises, available when needed and free from faecal and priority chemical contamination	Improved drinking water sources are those that, by nature of their design and construction, have the potential to deliver safe water. These include piped water, boreholes or tubewells, protected dug wells, protected springs, rainwater, and packaged or delivered water
Basic	Drinking water from an improved source, provided collection time is not more than 30 minutes for a round trip including queuing	
Limited	Drinking water from an improved source for which collection time exceeds 30 minutes for a round trip including queuing	
Unimproved	Drinking water from an unprotected dug well or unprotected spring	Unimproved groundwater and surface water sources are not assumed to be able to supply safe water consistently
Surface water	Drinking water directly from a river, dam, lake, pond, stream, canal or irrigation canal	

However, it is appropriate to commence this book with the individuals, households, and communities that are still left behind and dedicate it to those who are at risk of remaining excluded from the high ambition of the SDG target. In the final decade of the SDGs – and beyond – it will be necessary to find new energy and commitment to serve the hundreds of millions of people who are otherwise unlikely to enjoy safely managed services any time soon (Hutchings and Carter, 2018; Carter, 2019b).

About half a billion people in rural areas around the world still lack an improved water service – defined either as 'limited', 'basic', or 'safely managed' in JMP definitions (Table 1.1). Even water from improved water points has to be carried home on the heads or backs of children and women, often several times per day, so contributing to numerous health problems and injuries (Curtis, 1986; Geere and Cortobius, 2017). And yet such service levels represent a significant improvement on the use of unprotected groundwater sources or open surface water.

Rural and urban

It is a convenient simplification to divide people into 'rural' and 'urban' dwellers (see Annex). This book focuses on those living in rural areas. However, this category comprises a wide range, from those living in very remote and sparsely populated locations far from roads, towns, and cities (including nomadic and semi-nomadic pastoralists), those inhabiting dispersed or more compact village settlements, through to those living in trading centres and small towns (Figure 1.2). Furthermore, many urban people who live in small and large towns and cities keep in close touch with their rural origins, visiting family, maintaining their influence back home, sending money (remittances),

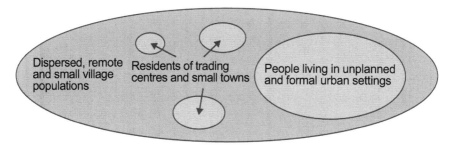

Figure 1.2 Simplified schematic showing spectrum of primary dwellings from rural to urban

and maintaining farms and houses in their rural communities. These connections, omitted from Figure 1.2 for reasons of clarity, include transfers of information, goods and services, and money within rural environments and between rural, small town and urban settings.

Rural water coverage

At the time of writing this book the most recent global and regional estimates of populations enjoying different service levels were for the year 2017. The estimates are shown in Table 1.2. Data on safely managed services is incomplete, so this category of service is omitted from the table.

The regions with the largest proportion of their rural populations lacking at least a basic service are Oceania and sub-Saharan Africa – both at around 55 per cent. In terms of absolute numbers, sub-Saharan Africa has about 337 million rural people who still lack at least a basic service; the two Asian

Table 1.2 Global and regional rural drinking water coverage, 2017, % and millions served

Region	At least basic		Limited		Unimproved		Surface water	
	%	Number	%	Number	%	Number	%	Number
World	81	2752	4	136	11	374	4	136
Australia & New Zealand	>99	4.1	<1	0	<1	0	<1	0
Central & Southern Asia	91	1130	1	12	7	87	2	25
Eastern & South-Eastern Asia	86	830	1	10	11	106	1	10
Europe & North America	98	249	<1	1.3	2	5.1	<1	1.3
Latin America & the Caribbean	88	114	2	2.6	6	7.8	5	6.5
Northern Africa & Western Asia	84	160	9	17	3	5.7	4	7.6
Oceania	44	3.9	2	0.2	6	0.5	48	4.3
Sub-Saharan Africa	45	276	17	104	25	153	13	80

Source: JMP, 2019a

regions contribute another 250 million; in total sub-Saharan Africa and Asia hold 90 per cent of the rural population still lacking at least a basic service. In total 640–650 million people still use surface water, unimproved groundwater sources, and so-called 'limited' services.

Getting it going and keeping it flowing

Improving physical infrastructure is the first stage, and the relatively easy part, of enhancing the water status of rural communities. In general, this stage is technically straightforward, cheap (although in many cases beyond the means of communities), and of short duration. Keeping the service working – the sustainability dimension – is much harder and costlier, and it requires continuing attention to management and financing aspects. Before coming to the sustainability challenges, however, the roles and responsibilities of water users, public and private sector organizations, and civil society need to be considered. Who invests in first-time improvements, and who manages services over time?

Self-supply involves the investment and management of improvements to services primarily by individual households (Sutton and Butterworth, 2021). There are numerous examples of such household-level initiatives, not only from low-income countries, but also from a wide range of wealthier economies including in Australia, Europe, and North America (RWSN, n.d., h). There will always be some households that are physically located beyond the reach of piped networks, but which can be adequately served by dedicated, self-managed services.

Community water supply refers to the service provided to part or all of a rural population by significant water supply infrastructure such as a protected spring, well or borehole, through to a piped system delivering water to public taps and house connections. It is uncommon for rural communities in low- and middle-income countries to invest in such systems without external financial help. Capital investments generally come from governments (with or without donor support) or from international NGOs. However, having made the capital investments, the norm is for new facilities to be 'handed over' to communities to manage and finance – a task that they may struggle to fulfil, especially as physical infrastructure ages and becomes increasingly costly to repair.

For some time now there has been interest in the potential of **private operators (for-profit or non-profit social enterprises)** to manage water services and so improve the reliability of supply. Capital investments may have been made by governments or NGOs, but by grouping and clustering such systems, and managing them together in a professional manner, performance may improve.

A fundamental difficulty with both community-managed and private operator-run services in low- and middle-income countries is the ability and willingness of water users to cover the full costs of providing the service. The national or international ambition to improve service levels exceeds the present ability of national economies and communities to finance the true costs; consequently, financial sustainability is a fundamental challenge (Carter et al., 2010; Franceys et al., 2016; McNicholl et al., 2019; Moriarty et al., 2013).

Attributes of drinking water services

Domestic water services have six main attributes, most of which are captured in the literature and discourse on the human right to water (UN-Water, n.d.). In all regions and countries, people value these attributes, but they take on varying degrees of importance to the extent that they are absent or compromised in any particular context.

Consumers want:

- a sufficient quantity of water;
- acceptable quality;
- they want to be able to take water when they need it – the supply should be reliable;
- convenience, especially through proximity;
- a service that is affordable;
- manageable services, given that they can only tolerate a certain level of difficulty.

In high-income countries, where most people enjoy safely managed services, potable water is on tap and on-demand in the home 24 hours per day, every day; water tariffs are relatively low; and the management functions that those tariffs pay for is undertaken by a professional utility. There is no question of community management. In this situation, using the human right descriptors, water is more than **sufficient** in quantity, **safe** to drink, **acceptable** in quality and delivery, physically **accessible** at multiple points in the home, and **affordable**. Notwithstanding occasional well-publicized incidents or accidents, and minor inconveniences caused by pipe bursts or pressure variations, the most common aspect that may lead to dissatisfaction is taste – perhaps explaining why some consumers choose to drink bottled water or filter their tap water, often at significant cost.

In cases where people do not enjoy even limited services (those using unprotected groundwater or surface water sources, or distant protected water points), most if not all of the six attributes of the service are compromised. Improved access and convenience – enabling greater consumption – commonly takes precedence for water consumers over water quality considerations. This is a rational choice by water users, as the benefits of using more water probably outweigh those of using better quality water, when prior levels of consumption are very low (Cairncross and Valdmanis, 2006).

For those using so-called basic services, the water source is protected and some form of community management arrangement usually exists. However, when community management becomes too difficult, repairs do not get done. When the costs of repairs become unaffordable, repairs do not get done.

When communities seek external assistance and this fails to materialize, repairs do not get done. The significant capital investments to provide physical infrastructure prove pointless as services deteriorate and fail.

Serving all, sustainably

This brings us to the two threads – the two imperatives – that run through this book. The first is to see all rural people having access to domestic water services that are progressively improving, until such time as they enjoy safely managed services. This is summed up in the idea of inclusion. As the SDG declaration put it, 'we pledge that no one will be left behind' (UN General Assembly, 2015). I do not argue for the least well served (those using surface water, unimproved groundwater sources, and limited services) to immediately receive a safely managed service; to propose that would lead to a few people becoming water-wealthy while the majority remain in water poverty. Progressive realization of the human right to water must mean that all are seeing improvements as they move up the water ladder of Table 1.1.

The second imperative, at least as important as the first, is that as rural people's water services improve, those services remain sustainable at every stage. This means that periods of non-functionality are few and of short duration. Breakdowns are inevitable, but it is the response of those accountable for diagnosing problems and carrying out repairs – be they communities, local governments, faith-based organizations, NGOs, for-profit private operators, social enterprises, or some combination of these – that really matters (Carter, 2019a). If services are not sustainable, then investments are wasted, people's well-being regresses, and it becomes ever more difficult to bring better services to all.

Inclusion is about serving everyone; sustainability is about keeping everyone's services working permanently. This book is about both.

Current trends and obstacles to progress

This book is written in the context of a number of important trends that have the potential to transform the lives of rural water consumers. First, there is increasing recognition that the full (so-called 'lifecycle') costs of improved water provision (not to mention safely managed services) are unaffordable to low-income consumers. The costs of keeping services working exceed the levels of revenues achievable through tariffs (water charges); sustainable service provision is only possible with subsidies from taxes (national budgets) and transfers (from foreign sources, official development assistance, and private philanthropic contributions).

Second, there have been some encouraging developments in recent years in technologies for accessing groundwater and improving water services. These include advances in technology for selecting suitable sites for well drilling, in drilling itself, and for lifting groundwater. Digital technologies for remotely monitoring the performance of water points and for facilitating payment of water tariffs offer opportunities to improve responsiveness of repair services and financial sustainability of management support organizations. Technology is rarely transformational on its own, but it can be part of the solution to intractable problems in rural water service delivery.

Third, the community management paradigm is coming under increasing challenge and critique at the time of writing, with the emergence of alternatives in the form of 'private operator' services (i.e. paid-for services, run by for-profit or not-for-profit organizations) at least in those places with sufficient population density and disposable income. It may not yet be time to discard the community management model entirely – in some places there are no viable alternatives; but it is important to consider alternatives too.

The book is written in light of a number of persistent obstacles to progress. These include failures in governance and political commitment among national governments; competition and failures in coordination by unaccountable external players; and failures of all the above to understand the barriers to participation commonly experienced by water consumers.

As a consequence, there is increasing recognition that national efforts to improve the rural water supply situation may require attention to numerous aspects of the 'system' and to the linkages between them. Systemic failures need to be addressed by concerted and patient work to address and strengthen weaknesses across the board.

The scope of this book

This book is about water services provided to rural communities by governments and their development partners (funders and NGOs) for drinking and other household uses. It highlights key aspects of groundwater resources, including pressures on the quantity and quality of those resources and the science and engineering of their exploitation. It addresses arrangements for managing and financing domestic water services, and the institutional and social aspects of rural community water supply. It recognizes differences between regions, and especially the transitions in technology and management that some regions have already made, some are making, and others are yet to make.

The focus in this book on groundwater and its exploitation is deliberate. Groundwater is water stored in aquifers, porous and permeable geological formations that lie below the earth's surface. Groundwater tends to be of better quality than surface water (although it is vulnerable to contamination (Lapworth et al., 2017). It can offer resilience in the face of drought and climate change (Cuthbert et al., 2019; MacDonald et al., 2019). Groundwater is the principal source of drinking water for communities in Africa and Asia (Margat and van der Gun, 2013; Velis et al., 2017; Carrard et al., 2019).

The main focus of this book is on the many rural communities who do not yet enjoy domestic water piped into the home, and who are unlikely to do so in the near future. The well-being of these communities could be significantly improved were they to have ready access to well-managed water points supplying

10 RURAL COMMUNITY WATER SUPPLY

water that is safe to drink, sufficient in quantity, dependable, affordable, and not imposing unrealistic management demands on them as users.

The statistics on access to drinking water show that around two-thirds of a billion rural people still do not have an engineered ('improved' in JMP terminology) water point within a 30-minute round trip (a so-called 'basic' service); of the three-quarters of a billion rural people who do enjoy a basic service, a common experience is of lengthy downtimes as a result of institutional, management and financing weaknesses. Another 2 billion rural people (mostly in the wealthier countries) enjoy treated water in the home, and it is this level of service that the UN rightly aspires to, as expressed in their Sustainable Development Goals, target 6.1.

In determining the scope of a book such as this, choices have to be made not only about the primary focus, but also about what to leave out (Figure 1.3). This book does not examine productive uses of water (for example in agriculture) except in passing references, for instance to the impact of agricultural abstractions on drinking water resources, or the relative ease or otherwise of managing domestic and productive water services. The subject of water in agriculture is

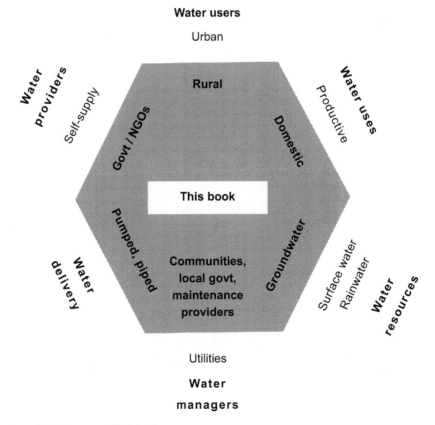

Figure 1.3 The scope of this book

the focus of a number of international agencies, many universities, and a few NGOs. Among the key international bodies are the UN Food and Agriculture Organization, the International Fund for Agricultural Development, the International Food Policy Research Institute, the International Livestock Research Institute, and the International Water Management Institute. The use of water for multiple purposes ('multiple-use services') is an important dimension of rural water supply, but it largely lies beyond what space permits in this book.

While recognizing that the distinction between rural and urban habitation is becoming increasingly blurred, this book generally does not address issues specific to urban water services by utilities; where it does so, it is to argue for some of the principles of utility-run services to be extended to rural water users, and vice versa.

This book says only a little about the initiatives that individuals and households can make to improve their water services. This important topic – self-supply – is comprehensively covered by Sutton and Butterworth (2021), so it is largely omitted here to avoid repetition.

Although the SDG target for domestic water is 'safely managed services', the reality is that many people will need to rely on non-networked point sources outside the home for many years to come. The priorities must be to improve water services for all, while simultaneously ensuring that those services are sustainable. Many hundreds of millions of rural people have yet to experience a safe and reliable service within a 30-minute round trip. A much larger number of rural people enjoy 'at least [a] basic' service, from a handpump-equipped borehole, a public tap in a communal piped water system, or, if they are particularly fortunate, a house connection; but their service is frequently far from dependable.

This book is about drinking water services provided by governments and their development partners (donors and NGOs) in places where 'safely managed' services are unlikely to be affordable for the foreseeable future. The themes running through the book are the two most important outcomes of water service provision – the sustainability of services and the imperative to 'leave no-one behind'.

Navigating this book

Chapter 2 presents an overview of the implications of water quantity and quality for the health and well-being of rural communities.

Chapters 3–6 provide an introduction to the main natural science and engineering aspects of rural water supply. Chapter 3 examines groundwater resources – both the quality and quantity aspects – and the pressures on those resources as populations grow and urbanize, and as climate change continues to bite. Good practices in regard to the exploitation and stewardship of groundwater resources are highlighted. Chapter 4 explores key aspects of borehole construction for performance and longevity. Emphasis is given to the implementation of known good practice, through transparent regulation and procurement, appropriate forms of contract, high-quality construction supervision and post-construction testing, and long-term monitoring. Chapter 5 provides an overview of water lifting, together with a review of the most commonly used rural water supply handpumps and the issues handpumps raise more broadly. Chapter 6 outlines the main infrastructure options 'beyond handpumps' – namely mechanised (including solar) pumping, and the main considerations in piped system design.

Chapters 7 and 8 introduce the management and financial aspects of rural water supply. Chapter 7 concerns the management functions needed to keep water services working, and the range of measures that are appropriate in different contexts. A plurality of arrangements is needed, from self-supply, through community management, to the involvement of faith-based organizations, private sector, and social enterprises. Chapter 8 is about the financial aspects of rural water services. Costs, ability, and willingness to pay are key aspects in designing user tariffs; the most difficult aspect, however, is the question of how the shortfall between revenues derived from tariffs and the costs of running services should be made up.

Turning to the human ('community') aspects of the subject, Chapter 9 examines the impacts of improved water services on rural people and their communities. It identifies those who benefit and those who may miss out in well-intentioned development interventions. Some principles for sound rural community water supply programming are identified.

In the final three chapters, I examine the challenges of rural water supply, and the new approaches that are being deployed to address these challenges. Chapter 10 asks why getting reliable, safe, and affordable water supply to rural communities is apparently so difficult, and it reviews various attempts to address the challenges. Chapter 11 examines 21st-century changes in approach, innovations in technology, and new ways of thinking about rural water services, which are current at the time of writing. As well as reviewing the 'systems strengthening' approach which is receiving increasing attention today, Chapter 12 invites you to imagine a different future in which sufficient resources are invested in rural water services, with particular focus on the requirements for sustainability; a future in which communities dependent on unprotected and limited services are 'levelled-up' so that all enjoy at least a basic water service in preference to investment in piped and safely managed services for the few; a future driven by the values and imaginations of professional and competent individuals and organizations working together for this common goal.

The contribution of this book

This book attempts to bring together aspects of the natural science, the social science, the engineering, and the management and financial dimensions of rural water supply under one cover. Its aim is to help guide governments, their funding partners, programme managers, and others concerned with rural water provision, during and beyond the last decade of the Sustainable Development Goals.

CHAPTER 2
Water quantity, quality, and health

Abstract: *This chapter focuses on those rural people – estimated to number about 1.6 billion globally – who do not enjoy a water service supplied at the home, many of whom will not experience the Sustainable Development Goal target of 'safely managed' water by 2030. Improving water access and increasing domestic water consumption, especially for home and personal hygiene, can establish one of the key prerequisites for reduction in diarrhoeal disease and neglected tropical diseases, including trachoma and schistosomiasis. While diarrhoea mortality in under-5s has fallen significantly in the last 30 years, the faecal-oral transmission route remains important. Recognition of environmental contamination by excreta of animal and human origin, environmental enteric dysfunction, the importance of food- and hand-hygiene, and the problem of stunting, all highlight the importance of drinking water quality and its relationship to sanitation and hygiene. Geogenic contamination of groundwater by arsenic, fluoride, iron, and manganese pose challenges for drinking water safety. Water carrying over long distances and rough ground has numerous health and safety implications, particularly for women and children. Poor rural community water services are compounded by inadequacies in service at schools and health care facilities. I argue for three overriding priorities: where domestic water sources are distant, improve access; where water is unprotected, apply basic engineering practices to safeguard water sources from contamination; and for all existing and future water services, focus on the social, scientific, organizational, managerial, financial, and technological aspects necessary to achieve sustainable and inclusive service provision.*

Keywords: water quantity, water quality, water carrying, water treatment, environmental enteric dysfunction, stunting, health, neglected tropical diseases, water service levels, sustainability

> '*Le mieux est l'ennemi du bien* (perfect is the enemy of good)'
> —Voltaire, 1764

Introduction

In this chapter I examine in greater detail the experiences of those rural people who for various reasons are not yet able to access water close to or in the home. These people generally rely on untreated surface water, unprotected and untreated groundwater, and 'improved' (i.e. protected but mostly untreated) groundwater. In most cases such water is collected from water points outside the home. In an increasing number of cases, spring water, surface water, and groundwater are delivered via piped systems, but many rural water consumers must still walk to a public water point, such as a tap or standpost. Getting safely managed water services – water on the premises, treated to safe standards, and on-demand – to these communities would be the best way to serve them, were it realistic and affordable. For many, it is not.

It is important to note too that supply of treated water via a piped system provides no absolute guarantee of good water quality. Failures in treatment processes and ingress of contaminated surface water into piped systems persist in many places.

The premise of this book is that the financial and management impediments to the achievement of the safely managed water target within the Sustainable Development Goal (SDG) timescale (everyone so served by 2030) mean that many people will still lack such a service even at the end of the SDG period. This being the case, the most constructive strategy is to progressively extend improved point water services to all, while simultaneously addressing the sustainability challenges associated with even these levels of service. Many of the disadvantages associated with the use of surface water, unimproved groundwater, limited, and even basic services – which are set out in the rest of this chapter – can be alleviated by addressing access and sustainability dimensions of the rural water problem.

Figure 2.1 illustrates the issue using Joint Monitoring Programme (JMP) data. In Central and Southern Asia access to safely managed services is rising at a rate that indicates the vast majority of the region's rural population may enjoy safely managed services by 2030. In sub-Saharan Africa, however, the prospects of more than 20 per cent of the rural population having safely managed services by 2030 are slight.

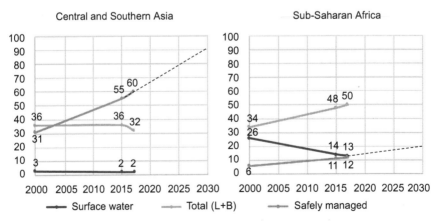

Figure 2.1 Trends in water services, 2000–17, Central and Southern Asia and sub-Saharan Africa, % of rural populations
Notes: L = Limited service; B = Basic service
Source: JMP, 2019b

Table 2.1 Rural domestic water requirements

Purpose	Daily amount per person, litres	Comments (those in quotation marks are from Howard and Bartram, 2003)
Drinking	1.0–5.5	Lowest values: children; highest: lactating mothers
Cooking	2.0	
Sub-total	**7.5**	Described as a 'basic minimum'
Personal and home hygiene	Implied 12.5	'Defining a minimum quantity of water is neither supported by evidence nor of practical value'
Sub-total	**20**	'A minimum for basic health protection'
Amenity uses	Variable	Not defined
Productive uses	Variable	Not defined
Grand totals	< 5	'No service'
	20	'Basic service'
	50	'Intermediate service'
	> 100	'Optimal service'

Water requirements and water consumption

Domestic water is needed for drinking, cooking, personal and home hygiene, and very small-scale productive uses. Howard and Bartram (2003) reviewed domestic water requirements and suggested a number of quantities and service levels, which are summarized in Table 2.1. The authors' proposal of four service levels in relation to water quantity, and the associated travel times and distances mentioned in the document, have been largely perpetuated through to the present-day SDG targets.

Two important points to bear in mind in all subsequent discussions of rural water supply are, first, that people need water for multiple purposes (the main ones of which are listed in Table 2.1); and second, that people use different water sources, either for different purposes (e.g. for drinking and for other domestic uses) or at different times (e.g. in wet and dry seasons).

Water quantity and health

Howard and Bartram's (2003) paper was framed in terms of health, in particular in relation to the consumption (drinking and cooking) and hygiene uses of water. A more recent review by Stelmach and Clasen (2015) concluded that in low- and middle-income countries using more water for face-washing of children can significantly reduce the prevalence of trachoma. Trachoma is an eye-disease caused by infection with the bacterium Chlamydia trachomatis; about 1.9 million people are blind or visually impaired as a result of the infection, and 142 million people are thought to be at risk in 44 countries (WHO, 2020b). Maintaining facial cleanliness is one of the four components of WHO's prevention and control strategy. The same review also found that increased use of water for personal and domestic hygiene was generally associated with reduced diarrhoeal disease.

Actual consumption

People who have to walk to a water point (be that a public tap or a groundwater source) make some sort of conscious or unconscious mental calculation as to how much time and effort to devote to this daily task; this has consequences for how much water they fetch. In general, the further the water point, the more time and energy are expended per journey and the less water is collected; but this relationship is not necessarily linear.

A frequently cited graph (originally published by Cairncross and Feachem, 1993) suggests that as a water point comes closer to the home and travel time shortens, consumption increases to a plateau of around 15 litres per person per day when travel time is less than 30 minutes. It is only when travel time is very short (equivalent to a tap or source in the compound or in the house) that consumption rises rapidly. The evidence for this was based on water-use studies in East, West, and Southern Africa, Nicaragua, India, Sri Lanka, and Bangladesh. More recent studies in Ethiopia (Tucker et al., 2014; MacDonald et al., 2019) provide support for the general form of the travel time/consumption relationship.

The quantity of water actually consumed at home is significant for health. In their study 'Drawers of Water II' Thompson et al. (2001: 99) concluded that 'once the consumption for drinking and cooking is satisfied by a limited and relatively invariable amount [about four litres per person per day], almost all the remainder is used for personal hygiene or cleaning utensils and house'. This is important, given the significance of hygiene for health (Curtis and Cairncross, 2003; Aunger et al., 2016). The authors do however qualify this conclusion with the remark that the actual health benefits of having more water available for household and personal hygiene may only be realized if practices such as hand-washing after defecation are promoted; they do not follow automatically. Water is needed in adequate quantities for menstrual hygiene, both at home and in school and other institutional settings.

In the SDG definitions of service levels, the difference between a 'limited' and a 'basic' service lies in the return-trip travel time. The source of data on travel times is the national household surveys, which supply information to the JMP (2018a). And yet evidence and common sense suggest that water users may have little idea of the exact time taken by their visits to water points. Probably no-one who fetches water in this way times their journey so specifically. The evidence from field studies on this matter questions the accuracy of self-reported time estimates. Both Ho et al. (2014) and Davis et al. (2012) find that these time estimates are approximate at best, putting in question the accuracy of the JMP statistics on basic and limited services (both of which were previously combined as 'improved' water services).

To summarize:

- In situations where travel times are very long (notionally more than 30 minutes) consumption of a few litres per person per day represents survival levels; maintenance of home hygiene (including keeping latrines clean) is

virtually impossible; personal hygiene and laundry are often undertaken at the distant water source, and handwashing after defecation or before food preparation or child feeding is unlikely to happen.
- Where round-trip travel times are in the notional range of about 5–30 minutes, a basic level of home hygiene is possible, but it cannot be assumed that hygiene practices automatically change to provide adequate safety in regard to faecal-oral disease transmission (see next section).

Looking ahead, there are strong arguments, both health-related and reflecting consumers' aspirations, for bringing water very close to, or into, people's homes; however, constraints of cost, and the added difficulties of keeping services working, mean that for many rural people this will remain a future (and distant) dream.

Water quality and health

Microbiological water quality and the faecal-oral transmission route

The connections between water quality (its 'safety') and health may seem both obvious and critical to human well-being. However, there can be a tendency among water professionals to elevate the importance of drinking water quality, sometimes to the exclusion of other aspects of water services. This is not to say that drinking water quality is inconsequential, but that there are nuances and contextual issues that need to be considered when determining the importance to place on this subject.

It is generally accepted that those most at risk from unsafe water quality are the young. The first 1,000 days of life (from conception to 24 months) are critical in terms of child development, and this recognition has been enshrined in the BabyWASH Coalition (SDG Partnership Platform, n.d., b). The ingestion of faecal pathogens – in part delivered via unsafe (microbiologically contaminated) drinking water – may lead to acute infectious and chronic diarrhoea, which in turn can result in malnutrition, poor cognitive development, and stunting.

The faecal-oral transmission route of disease has been summarized in the well-known F-diagram, first produced by Wagner and Lanoix (1958). Others have updated the diagram in recent years (Penakalapati et al., 2017; USAID, 2018; Budge et al., 2019), in particular to reflect recent thinking about the importance of animal excreta as a source of diarrhoea-causing pathogens and the direct consumption by infants of contaminated soil (known as geophagy) (Figure 2.2).

The F-diagram makes clear that disease-causing bacteria, viruses, and protozoa (pathogens), which are present in human and animal excreta, can enter the mouths of young children (and adults) by a variety of routes. Drinking contaminated water ('fluids') is only one of these routes, and it is not necessarily the most important. Babies and infants can ingest pathogens from the ground on which they sit and play ('fields' and 'floors'), from their

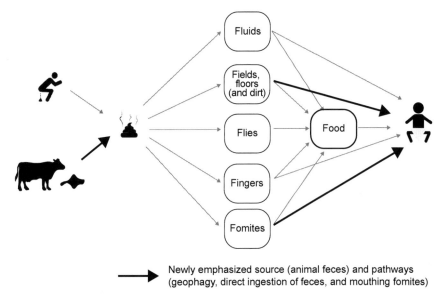

Figure 2.2 A simplified form of the F-diagram, incorporating animal excreta
Source: USAID, 2018, used with permission; this version is derived from figures in Penakalapati et al., 2017

own hands ('fingers'), from the hands of those who feed them, from food, and from surfaces ('fomites'), which harbour microorganisms. In recent years, recognition of the importance of hands and food, as well as animal excreta, as risk factors has been growing. Cairncross and Valdmanis (2006) stated that 'five types of evidence suggest that domestic hygiene—particularly food and hand hygiene—is the principal determinant of endemic diarrheal disease rates and not drinking water quality'. This conclusion is still relevant in situations where people largely live outdoors, where indoor and outdoor environments are contaminated with animal and human excreta, and where the maintenance of household and personal hygiene is difficult.

The presence of disease-causing pathogens in drinking water is not usually determined directly (e.g. by attempting to measure the presence of cholera-causing bacteria), but rather indirectly by detecting and counting faecal indicator bacteria, in particular the thermotolerant coliform bacteria (also referred to as 'faecal coliforms' and loosely identified as Escherichia coli). The World Health Organization (WHO) is of the view that such bacteria, and the pathogens that they are taken to indicate, should be absent from drinking water. In a systematic review of microbiological water quality, Bain et al. (2012: 925) concluded that 'microbial contamination is widespread in lower- and middle-income countries and affects all water source types, including piped supplies. Drinking water is more likely to be contaminated in rural areas than urban areas, and faecal contamination was most prevalent in Africa and South-East Asia'. More recent work (Lapworth et al., 2020) in Ethiopia,

Malawi, and Uganda, however, has concluded that the problem of microbial contamination of water delivered by rural handpumps may sometimes be less widespread than these conclusions would suggest.

In order to protect water sources and resources from contamination by pathogens, sanitary surveys (Pond et al., 2020) and water safety plans (Bartram et al., 2009) are often considered to be useful.

The health implications of deficiencies in water services are not limited to those caused by the faecal-oral transmission route. A range of other diseases can be linked, directly or indirectly, to deficiencies in water supply.

Environmental enteric dysfunction (EED)

EED is a damaging condition of the small intestine lining, thought to arise from ingestion of high loads of human and animal faecal microbes (Korpe and Petri, 2012; Crane et al., 2015). It inhibits nutrient uptake, so leading to chronic undernutrition and stunting. Drinking water represents only one of multiple sources of faecal microbes, ingestion (especially by young children) of contaminated soil and food probably being of greater significance (Ngure et al., 2013).

Stunting

Impaired growth resulting from poor nutrition, especially in the first 1,000 days of life, is referred to as stunting. Stunting has consequences in childhood and adulthood, including poor cognitive development and educational attainment, low income, and later nutrition-related chronic disease.

The problem of stunting lies at the interface between nutrition and environment; low height for age is thought to be the result of undernutrition caused directly by food insecurity or indirectly by EED – in part exacerbated by inadequacies in water, sanitation, and hygiene (WASH) – or both.

Ingestion of microbiologically contaminated water represents one possible component of a much more extensive and complex picture of the causes of stunting.

Based on the hypothesis that both childhood diarrhoea and stunting may result from a combination of inadequate WASH together with poor nutrition, two recent randomized controlled trials – the WASH Benefits trial in Bangladesh and Kenya (Luby et al., 2019; Null et al., 2019) and the SHINE trial in Zimbabwe (Humphrey et al., 2019) – were undertaken. In both cases, the tested intervention to improve water services focused only on the microbiological quality of drinking water. In neither trial was any measured impact found linking improved water quality (or indeed any other aspect of WASH) to stunting.

Neglected tropical diseases

WHO (2015) has identified 17 neglected tropical diseases (NTD) that together affect about 1 billion people. It set out a roadmap for the 'control, elimination and eradication' of these diseases (WHO, 2012), a guide (WHO, 2018), and a strategy linking WASH to progress towards that goal (WHO, 2015). An updated roadmap aligning with the SDG end-date of 2030 was published in 2020 (WHO, 2020b).

As far as rural water is concerned, the key aspects highlighted by WHO are coverage, access, use, safety, sustainability, and functionality. In particular the WASH and NTD strategy argues for the following actions (in addition to others that are more directly focused on sanitation):

- Prevent consumption of contaminated water, reduce contact with surface water, and enable personal hygiene practices.
- Water resource, wastewater, and solid waste management for vector control and contact prevention.
- Hygiene measures such as handwashing with soap, laundry, food hygiene, face washing, and overall personal hygiene.
- Availability of water for facility-based care and self-care (especially leprosy and lymphatic filariasis).
- Hygienic conditions for surgical procedures (e.g. for lymphatic filariasis hydrocele and trachoma trichiasis surgeries).
- Accessible water and sanitation services for individuals with physical impairments and caregivers.
- Measures to prevent stigma-based exclusion from water and sanitation services, including measures to enable personal hygiene and dignity.

Reduction of direct contact with surface water bodies that may be instrumental in transmission of schistosomiasis, and close and unrestricted access to enough water to maintain good hygiene – so limiting transmission of trachoma and soil-transmitted helminth infections – are key factors.

Drinking water chemistry

Several inorganic (Figure 2.3) and organic chemical constituents of untreated drinking water can have adverse health impacts. To put that statement in context however, WHO (2017a: 156) states that while 'few chemical contaminants have been shown to cause adverse health effects in humans as a consequence of prolonged exposure through drinking-water ... this is only a very small proportion of the chemicals that may reach drinking-water from various sources.'

Although treatment of water derived from community water points, either at the water point or in the home, is technically possible, the practice of water treatment on these limited scales is less straightforward than for piped services (see following section).

The subject of drinking water quality is extensive, and only a brief overview of key issues can be given here, by reference to the main chemical constituents that affect health. More extensive coverage of the subject is given in Freeze and Cherry (1979), Fetter (1994), MacDonald et al. (2005), and Misstear et al. (2017).

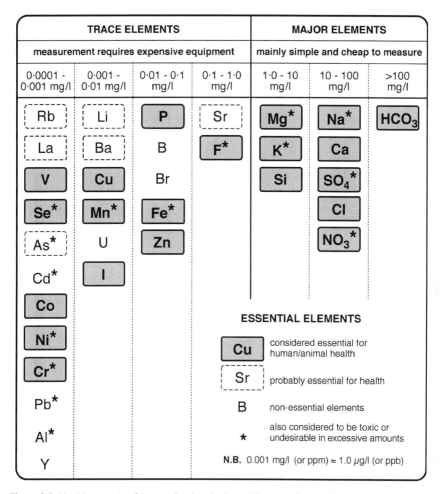

Figure 2.3 Health aspects of inorganic chemical constituents of groundwater
Source: MacDonald et al., 2005, adapted from Foster et al., 2000, and Edmunds and Smedley, 1996

From a health perspective arsenic and fluoride are the two most important inorganic species in the rocks that contain groundwater. Because of their origins in the earth's crust they are referred to as geogenic contaminants (Bader et al., 2017). In the following paragraphs, reference to WHO limits are derived from the Fourth Edition of the WHO's 'Guidelines for drinking water quality' (2017a).

Arsenic (Ravenscroft et al., 2005; Appelo, 2006) was first found as an important, widespread, and harmful constituent of groundwater-derived drinking water in Bangladesh in the early 1990s. It has subsequently been detected in numerous other regions – North and South America, Europe, Africa, and Asia.

Intake of arsenic may take place from drinking water or food, with rice and other cereals contributing significant proportions of the daily intake. Cumulative intake of arsenic over many years can lead to 'various health effects including skin problems (such as colour changes on the skin, and hard patches on the palms and soles of the feet), skin cancer, cancers of the bladder, kidney, and lung, and diseases of the blood vessels of the legs and feet, and possibly also diabetes, high blood pressure, and reproductive disorders' (WHO, n.d.). WHO has set a safe limit on arsenic concentration in drinking water of 10 micrograms per litre.

The relative proportions of daily arsenic intake from drinking water and food are likely to be very case-specific. Daily quantities of drinking water vary, cooking methods (of rice especially) vary – in some cultures surplus water being used and in others all the cooking water being absorbed by the rice; and diets vary. In a study in rural Bangladesh (Ohno et al., 2007), in an arsenic contaminated area, drinking water was found to represent only 13 per cent of daily arsenic intake. However, the authors noted that (a) people avoided known arsenic-contaminated drinking water sources, while (b) there was evidence that they still used high-arsenic water for cooking. Each situation is likely to be different.

Exposure to excess arsenic intake is thought to put around 140 million people in 70 countries at risk (Ravenscroft et al., 2009).

Fluoride is common in areas underlain by volcanic geology and is present in concentrations that are damaging to health in locations such as the East African rift valley, but also in other geologies (including crystalline rocks containing fluorine-rich minerals, shallow aquifers in arid areas experiencing strong evaporation, sedimentary aquifers undergoing ion exchange, and locations with inputs of geothermal water) in all world regions (BGS, n.d.).

Fluoride intake is necessary in moderate amounts for dental protection, but beyond the safe limit set by WHO (1.5 milligram per litre) water consumers may experience dental fluorosis (damage to, and staining of, the teeth) and in extreme cases skeletal fluorosis (damage to bones, joints, and muscles). It has been estimated that more than 200 million people drink water with a fluoride concentration exceeding the WHO guideline value.

Iron is a very common constituent of groundwater. Other than in very high concentrations it does not pose a health hazard, and WHO sets no health-related limit on its concentration in drinking water. Its effect is indirect. Above a concentration of about 0.3 milligrams per litre it causes rust-coloured staining of laundry and cooking utensils, and it affects the taste of drinking water and cooked food. In the search for alternatives, water consumers may resort to sources that are far inferior in microbiological terms and from a health perspective.

Manganese is often found together with iron in groundwater, and WHO has set a health-related limit on its concentration of 0.4 milligrams per litre. Like iron, manganese imparts an unpleasant taste to drinking water and it leads to black staining of items that come into contact with it. Homoncik et al. (2010) have linked a number of neurological conditions to elevated manganese concentrations, but they stress the need for further research regarding acceptable health-related limits.

Nitrate concentrations in rural groundwater are generally low, the main exception being in areas that are intensively farmed. Elevated nitrate concentrations are also found in groundwater that is contaminated by human and animal faecal waste. The main health risk is to bottle-fed infants who may develop the rare condition methaemoglobinaemia – which prevents binding of oxygen to haemoglobin in the blood – and thyroid effects.

A wide range of **organic compounds** originating in agriculture (pesticides and herbicides), industry (industrial chemicals and hydrocarbons), and pharmaceuticals (medicinal drugs) pose increasing threats to groundwater quality globally. However, the main impacts of such contamination lie in industrialized urban and intensively farmed agricultural areas. Their further consideration lies beyond the scope of this book.

Water treatment

It is technically feasible to undertake water treatment for the removal or reduction of both microbiological contamination and some undesirable chemical constituents. Examples at community level include arsenic removal (Nicomel et al., 2016), fluoride removal (Mohapatra et al., 2009), and multi-stage sand filtration (Visscher, 2006). Treatment at the point of collection is mostly focused on chlorination (Kremer et al., 2011). At household level a range of water treatment methods is available (Hunter, 2009; WHO, 2016) mainly for the reduction of microbiological contamination. The Kanchan water filter, designed in Nepal by Ngai et al. (2007) effectively reduces both microbiological constituents and arsenic.

In recent years modern water treatment technologies such as membrane filtration and ultra-violet treatment have been deployed; however, the cost and management implications of such methods have often proved challenging.

The effectiveness of water treatment methods relies on consistent, correct, and sustained management. In many of the contexts that form the focus of this book, however, management and financial impediments severely constrain the possibility of such proficient management and consequent performance.

Water treatment in rural areas is relatively uncommon, and where it is practised, effective management is even less common. For example, in a study of 67 countries Rosa and Clasen (2010) estimated that 33 per cent of households reported some form of household water treatment. Of these, half reported simply straining water through a cloth or allowing it to stand and settle – neither practice rendering water 'safe' to drink. The practice was more common in urban than rural areas, and less common among poorer wealth quintiles. The practice is more common in Asia than in Africa.

WHO supports an international network – the Household Water Treatment and Safe Storage (HWTS) Network – devoted to the promotion of household water treatment and safe storage technologies and practices.

Recontamination

Numerous studies (for example Trevett et al., 2005a, 2005b) have shown that water of (relatively) good microbiological quality at source (or after treatment) may become contaminated during collection and transport, or in the home. Local contamination of groundwater may occur because of the absence of adequate sanitary seals (see Chapter 4), while contamination after collection can occur in transport or storage. There has been debate about the significance of recontamination for health, with some (e.g. Feachem, 1978; VanDerslice and Briscoe, 1993) playing down its importance on the grounds that pathogens, including those introduced into otherwise clean water, are in any case shared in the home with those derived from food and other sources; while others (WHO, 1999; Trevett et al., 2005a) argued against this view on the grounds that no household is an island, somehow isolated from those around it. Whichever view is taken, the general consensus today is that water collected and carried home should be transported in clean, closed vessels and stored safely in the home to avoid recontamination – although the 'safe storage' aspect of 'household water treatment and safe storage' is generally less well researched and documented than the treatment aspect (however see CDC, 2011, for an exception).

The implications of water-carrying

A number of studies of the burden of water carrying have been undertaken since the early publication by Curtis (1986). Geere and Cortobius (2017) made a broad review of this subject, examining the proportion of households that have to rely on 'off-plot' water sources, who within the household undertakes water collection, and the amount of time spent on this task. These authors and others (Pickering and Davis, 2012; Geere et al., 2018) have evaluated the health impact of daily water carrying.

The implications of having to carry water home from a water point some distance away include the following:

- The direct physical and musculoskeletal damage caused by carrying heavy loads over rough ground, especially in the case of children and pregnant women (Geere et al., 2018).
- The time and energy expended, which often keeps children out of school and contributes to fatigue and possibly depression (Cooper-Vince et al., 2017).

- The problems encountered by elderly and disabled people who may not be physically able to carry such loads (Jones and Wilbur, 2014).
- The risks of physical and sexual violence, particularly for girls and women (House et al., 2014).
- The breakdown of trust and potential for violence when men are suspicious of women's long absences.

All of these implications support the strategy of bringing water closer to home – ideally into the home. Until that is realistic and feasible, it is imperative to move people up the water service level ladder set out by JMP (see Chapter 1).

Water in rural schools and health facilities

It is not enough to focus only on water services at home when key public locations such as schools and health facilities, as well as council offices, markets, prisons, and other places, are deficient in this regard. The Joint Monitoring Programme commenced monitoring the state of WASH in schools and health facilities at the outset of the SDG period, with the first major baseline reports following in 2018 and 2019 (JMP, 2018b, 2019b).

Schools

The service levels applicable to WASH services in schools mimic those for community settings. Four levels are envisaged. The sanitation and hygiene descriptors are included here (Table 2.2) for completeness, to emphasize the need for water in maintaining cleanliness of sanitation facilities, as well as for general hygiene and menstrual hygiene requirements.

Table 2.2 JMP service ladders for WASH in schools

Level of service	Drinking water	Sanitation	Hygiene
Basic	Drinking water from an improved source and water is available at the school at the time of the survey	Improved sanitation facilities at the school that are single-sex and usable (available, functional, and private) at the time of the survey	Handwashing facilities with water and soap available at the school at the time of the survey
Limited	Drinking water from an improved source but water is unavailable at the school at the time of the survey	Improved sanitation facilities at the school that are either not single-sex or not usable at the time of the survey	Handwashing facilities with water but no soap available at the school at the time of the survey
No service	Drinking water from an unimproved source or no water source at the school	Unimproved sanitation facilities or no sanitation facilities at the school	No handwashing facilities or no water available at the school

Source: JMP, 2020a

26 RURAL COMMUNITY WATER SUPPLY

In the JMP baseline report (JMP, 2018b), the evidence shows that in 2016:

- In all but one region (Australia and New Zealand), some schools effectively had no water service; Oceania and sub-Saharan Africa were worst served in this respect, with 44 per cent and 47 per cent of schools respectively having no service.
- In central and southern Asia, about one-third of schools had an improved (but not necessarily working at time of survey) water service; two-thirds had a basic service.
- The regions with the lowest levels of basic sanitation and hygiene services were Oceania and sub-Saharan Africa; central and southern Asia were next worst served.
- Disparities in service level exist between educational level (schools serving the youngest tend to have the poorest services) and between rural and urban locations (rural schools being less well served).
- Provision of facilities for menstrual hygiene management vary a great deal, but this remains an aspect needing a great deal more attention.

Health care facilities

As in schools, JMP has defined new service level descriptors for WASH in health care facilities (Table 2.3). The data on WASH in health care facilities is even more limited than that for WASH in schools. However, the evidence suggests that:

- Sub-Saharan Africa has the lowest proportion of its health care facilities having a basic water service; in the same region about one-quarter of facilities have a limited service and another quarter have no service.
- Government facilities and rural areas are the least well served.

An emerging issue – antimicrobial resistance

The overuse of antibiotic drugs globally (but especially in low-income countries) has resulted in increasing numbers and variety of antibiotic-resistant bacteria entering the environment via human excreta and wastewater. Human exposure to such bacteria and related genetic material represents a significant and growing health risk (WHO, 2014; Fink et al., 2019). A detailed analysis of the matter is beyond the scope of this book, but it is clear that it will have growing implications in future for the linkages between consumption of untreated water and health, and regarding strategies for water treatment.

Progression up the service level ladder

Surface water users

Globally, about 136 million rural people claim to use surface water as their main source of drinking water. In absolute terms, the majority (about 78 million) live in sub-Saharan Africa, with another 34 million in Asia. By proportion,

Table 2.3 JMP service ladders for WASH in health care facilities

Level of service	Water	Sanitation	Hygiene	Waste management	Environmental cleaning
Basic	Water is available from an improved source on the premises	Improved sanitation facilities are usable, with at least one toilet dedicated for staff, at least one sex-separated toilet with menstrual hygiene facilities, and at least one toilet accessible for people with limited mobility	Functional hand hygiene facilities (with water and soap and/or alcohol-based hand rub) are available at points of care, and within five metres of toilets	Waste is safely segregated into at least three bins, and sharps and infectious waste are treated and disposed of safely	Basic protocols for cleaning are available, and staff with cleaning responsibilities have all received training
Limited	An improved water source is within 500 metres of the premises, but not all requirements for basic service are met	At least one improved sanitation facility is available, but not all requirements for basic service are met	Functional hand hygiene facilities are available either at points of care or toilets but not both	There is limited separation and/or treatment and disposal of sharps and infectious waste, but not all requirements for basic service are met	There are cleaning protocols and/or at least some staff have received training on cleaning
No service	Water is taken from unprotected dug wells, springs, or surface water sources; or an improved source that is more than 500 metres from the premises; or there is no water source	Toilet facilities are unimproved (e.g. pit latrines without a slab or platform, hanging latrines, bucket latrines) or there are no toilets	No functional hand hygiene facilities are available either at points of care or toilets	There are no separate bins for sharps or infectious waste, and sharps and/or infectious waste are not treated/disposed of safely	No cleaning protocols are available and no staff have received training on cleaning

Source: JMP, 2019b

Oceania is the most extreme, with 48 per cent of its rural population using surface water as their main drinking water source.

For those living near perennial surface water (lakes and reservoirs, large and small rivers, and irrigation canals) these offer convenient, reliable, and affordable water. It has been estimated (Kummu et al., 2011) that about 50 per cent of the world's population live within 3 km of such surface water, and between 5 per cent and 10 per cent within 1 km.

Exposure to high-risk faecal contamination and, in endemic areas for schistosomiasis, to the chronic disease that can result means that alternatives need to be found.

The priorities for rural people using surface water as their main domestic water source must be to enable access to water points that are closer to home and protected from contamination. In many if not most cases this will involve access to groundwater.

Users of unimproved groundwater

Globally, about 374 million rural people are thought to rely on unimproved groundwater sources (unprotected springs and wells). Water access and water quality considerations mean that for these water users adequate source protection and construction of new sources must be priorities.

Users of protected springs and wells

Globally, about 1,087 million (just over 1 billion) rural people use improved water points (classified as a limited service if the round trip time exceeds 30 minutes, and a basic service if it lies within 30 minutes). For these water users, water quality risks are less pressing, but an important priority is for more water points closer to home.

For all water point users

In addition to the safety and access priorities outlined above, the key imperative is to ensure high levels of functionality and up-time; to assure a sustainable service, a set of social, scientific, organizational, managerial, financial, and technological issues need to be addressed. These form the subject matter of the remaining chapters.

CHAPTER 3
Groundwater resources

Abstract: *The vast majority (98–99 per cent) of the earth's store of fresh liquid water is groundwater. Large volumes of groundwater can mitigate periods of drought, though this is dependent on renewable freshwater resources. This chapter outlines the land-based water cycle as well as the water balance concept, which are fundamental to the understanding of water resources. Depletion of groundwater, especially by over-abstraction for irrigated agriculture, has adversely affected rural drinking water supplies in many regions of the world. Although accurately estimating recharge – determined by climate, soil properties, and land use – is inherently difficult, understanding its magnitude is key to knowing the limits of sustainable abstraction. In addition, understanding the hydraulic properties of aquifers, particularly with regards to groundwater storage and flow, can indicate limits on abstraction rates for individual wells and boreholes. Determining the aggregate sustainable yield of an aquifer thus requires understanding how all parts of the water balance will change in response to pumping. These factors are increasingly impacted by population growth and climate change, which place greater stresses on water resources.*

Keywords: water resources, groundwater, water cycle, water balance, aquifers, groundwater hydraulics, climate change, groundwater recharge, groundwater abstraction

'Groundwater: our hidden treasure'
—Title of UNESCO conference, 2015

Introduction

The water resources on which rural water supplies depend must be sufficient in quantity and appropriate in terms of water quality for that purpose. A number of key aspects of water quality were addressed in Chapter 2, together with further references to more complete coverage of that important topic. This chapter provides an overview of important issues that determine the quantities of groundwater available for supply in rural areas. It is about the physical science of the resource on which engineered water supply systems and human management institutions depend to provide sustainable rural water services.

The quantitative aspect of water resources consists of two important elements: the amounts of groundwater in storage; and the quantities of water entering, flowing through, and leaving those groundwater stores. Both need to be considered when evaluating the quantities of water available for various uses. Stores of groundwater far exceed those stored anywhere else in the water cycle. Groundwater has been estimated to represent 98–99 per cent of all liquid freshwater on earth (Margat and van der Gun, 2013), although not all of this is readily accessible. The same authors estimate the amount of groundwater in storage globally as 8–10 million km^3. In separate work MacDonald et al. (2012) estimated the groundwater store in Africa as 0.66 million km^3. Such colossal figures are relevant to the extent that they indicate the buffering capacity of the sub-surface – the ability these stores of groundwater have to allow the continuation of abstractions (i.e. extraction of water) through dry periods – but they do not alone provide a measure of the amounts that can be sustainably withdrawn for use.

Equally important are the flows – the renewable freshwater resources – defined by FAO (2003) as 'the long-term average annual flow of rivers (surface water) and groundwater'. National estimates of renewable fresh water resources are available via the Aquastat database (FAO, n.d.). The total annual flow of fresh groundwater is a significant figure – about 2000 m^3 per person per year (Döll and Fiedler, 2008), of which only a few per cent is required for domestic use. The same authors estimate that renewable groundwater resources globally are about one-third of total renewable water resources. The total volumes and the proportions of renewable surface water and groundwater are highly variable spatially as a consequence of variable climates and geology.

Water flows and stores matter, both in nature and in engineered water supply systems. If abstractions of water continuously exceed the long-term renewable resources, then damaging impacts on natural discharges and abstractions from shallow wells may be experienced. The quantities of water in storage become progressively depleted, and such abstractions may therefore be deemed unsustainable.

In the following sections I focus on key aspects of water resources, inasmuch as these determine how water resources can be developed, protected, and managed for rural water supply. Water resources are dynamic, varying over seasons, from year to year, and over longer time scales.

The water cycle and water balance(s)

The water cycle

In the earth's environment, water forms one of the important natural cycles (Figure 3.1) as all human beings must depend on it for survival. Water falling to earth (correctly referred to as precipitation) may be in the form of rainfall, fog, and dew; and in its solid forms as hail, sleet, and snow. As space is limited, the remainder of this chapter considers water resources to start with rainfall, as the hydrology of snow and other solid forms of precipitation is a more specialized and location-specific topic (such as in those south Asian countries supplied by Himalayan snowmelt).

Rain falling to earth may follow, in varying proportions, three main possible routes. It may be intercepted by vegetation, and subsequently make its way to the soil surface or evaporate. It may infiltrate into the soil, most of it subsequently evaporating or being taken up by the roots of vegetation,

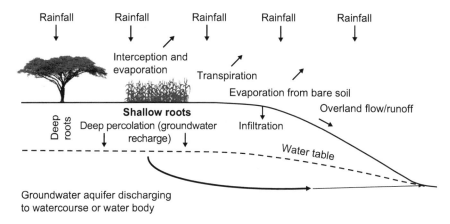

Figure 3.1 A simplified version of the water cycle
Note: 'Rainfall' in the figure also includes other forms of precipitation; deeper aquifers may also exist, separated from the unconfined aquifer shown, by a confining (low permeability) layer

and transpired back to the atmosphere. Part of it may form shallow subsurface flow or overland flow to watercourses and water bodies.

Some of the water that has entered the soil may continue to percolate slowly downwards to an underlying groundwater store known as an aquifer, the upper boundary of which is the water table. Having reached the aquifer, the natural tendency is for groundwater to move slowly towards watercourses and water bodies, supplementing their flows. All water that has evaporated or transpired to the atmosphere is, in principle, available to be recycled as precipitation, either locally or at more distant locations. Rivers discharge to the sea, from which water evaporates back to the atmosphere. And so the hydrological or water cycle continues.

Significant quantities of water may be made available for rural water supply from direct capture of rainfall and by abstraction of surface water and groundwater. The proportion of the total rain falling on a large area that can realistically be captured directly is small, although it can be very significant to those who can make use of it. Surface water – water bodies including wetlands and lakes, and watercourses (rivers and streams) – is the most visible component of exploitable water resources, but likely to be the most heavily contaminated. However, groundwater – the invisible resource – is especially important for rural water supply because of its near ubiquity, generally good quality, and because of the large volumes in storage, which can maintain supply even through dry periods.

The land-based water balance

In their focus on water resources, the sciences of hydrology and hydrogeology are directed to understanding in detail the processes within the water cycle, and to quantifying the various flows and stores. The relative magnitude of

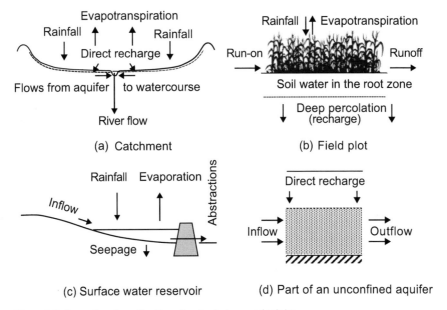

Figure 3.2 Examples of application of water balance principles

the various water cycle components varies enormously from place to place, and over time, depending on climate and weather, soil properties, vegetation cover, topography, and underlying geology.

The concept of the water balance is fundamental to water resources studies. It can be applied to any part of the natural or man-made system that may be of interest, including a catchment or experimental field plot, a surface water body, a reservoir, or an aquifer (Figure 3.2). The water balance principle is simple: inputs of water are exactly balanced by the sum of outputs and changes in the amount of water in storage.

When applied at the scale of a catchment (the geographical area within which rainfall makes its way to a river system), the input over a chosen period is the rainfall. The outputs are the river flows leaving the catchment, evaporation from bare soils and exposed water surfaces, transpiration from vegetation, and deep percolation or recharge to the underlying aquifer. Changes in storage take place in the soil and in open water bodies as their water content vary over time.

A key aspect of the water balance is that all the components are linked. Consequently it would be misleading to consider a change in one component without considering how this affects all the others. One important example of this is the impact of changes in rainfall on groundwater recharge. Under climate change, it is likely that rainfall will become more concentrated into high-intensity storms, regardless of whether the total amount at a particular location increases or decreases in the future. A trend toward higher-intensity rainfall events, even with a reduction in mean annual rainfall, may lead to more recharge, not less (Cuthbert, Taylor et al., 2019).

A second example is the well documented 'Sahel paradox', in which in certain areas of west Africa, declining rainfall over many years – thought to have been the largest and longest observed rainfall reduction seen anywhere on the planet – was accompanied by rising (not falling) water tables and increased (not decreased) river flows (Descroix et al., 2009). The main explanation was found in the fact that as population pressure grew, natural deep-rooted vegetation was progressively cleared, being replaced by annual crops with smaller water demands. The change in land use was enough to outweigh the drying tendency of the climate.

An important way in which hydrological sciences are pursued is through the use of models. Models are simplifications of reality, in which understanding of processes and their interactions form the basis for their quantification. A good model (judged by its ability to represent the past performance of the system with reasonable accuracy) can be used to explore possible futures by asking 'what if?' questions. What if rainfall increases (or decreases)? What if different land uses are implemented? How might changes in such variables affect river flows or groundwater recharge? At a time when major changes are taking place in terms of climate and demography, with knock-on effects on the natural and man-made environment, the ability to model possible futures is important. Water balance considerations form a key foundation of all water resource models.

The water balance concept is therefore of importance for at least three key reasons. First, it emphasizes the inter-connectedness of the entire water cycle, so urging caution in drawing simplistic conclusions about the impacts of, for example, climate change on water resources. Second, it underpins all the applied theory of groundwater flow and provides a fundamental principle in the modelling of groundwater resources. Third, in the right circumstances it forms the basis of an important approach to the estimation of groundwater recharge (Lerner et al., 1990; Rushton et al., 2006; Healy, 2013).

Components of the water cycle

Rainfall

Rainfall totals and averages conceal at least as much as they reveal, due to high variability over space and time. In a single location, two years may have similar rainfall totals, but many aspects of the year's rainfall may have differed between the years – the start and end date(s) of the rainy season(s), the number of rain-days or rainfall events, the amount of rainfall per rain-day or rain event, the duration of each rainfall event, and so on. All of these variables have implications for the water balance, and therefore for water resources.

Furthermore, annual rainfall totals often vary considerably from year to year, with drier climates often experiencing the highest variability (Figure 3.3). Rainfall can also vary greatly from location to location, even over quite short distances. This is especially true in dry regions and in places with pronounced relief.

34 RURAL COMMUNITY WATER SUPPLY

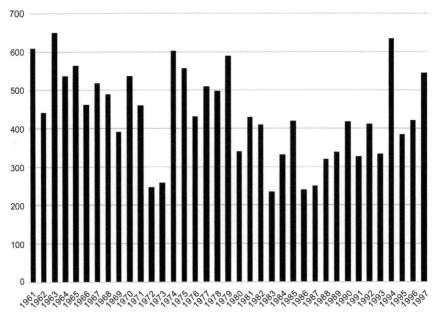

Nguru, north-east Nigeria, a semi-arid location. Mean annual rainfall 436 mm, CV 27%

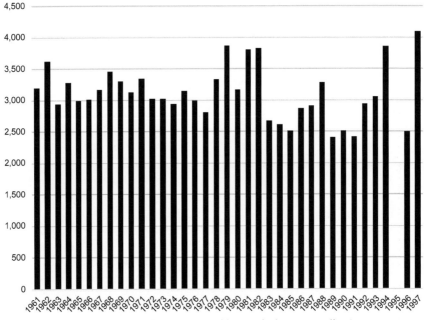

Makeni, northern Sierra Leone, with a tropical monsoon climate.
Mean annual rainfall 3,111 mm, CV 14%

Figure 3.3 Rainfall variability from year to year, 1961–97
Note: There are differences in vertical scales. No data available for Makeni in 1995.

The degree of variability is described by the coefficient of variation (CV): the higher the value, the more variable the data. Runs of drier-than-average or wetter- than-average rainfall are common, and so a true picture of what is 'normal' may require many decades of data. Climate change makes that determination even more complicated. In short, it is crucial to bear in mind that rainfall patterns are neither simple nor fixed; on the contrary, rainfall exhibits many different properties (not merely the mean annual value), and these can vary considerably over time and space.

Evaporation and transpiration

Second to rainfall, the total amounts of water evaporated and transpired back from the earth's surface to the atmosphere usually represent the next largest component of the water cycle. Both evaporation (from bare soil and open water) and transpiration (from vegetation) involve the conversion of liquid water into water vapour. Both are driven by the same physical processes. Consequently they are often lumped together and referred to as evapotranspiration (often abbreviated to ET).

Evapotranspiration rates are largely determined by weather variables. At a particular location, if any or all of solar radiation, temperatures, and windspeeds are high, and/or relative humidity is low, the higher is the rate of evapotranspiration. Evapotranspiration (more strictly potential evapotranspiration) is usually estimated from one or a combination of these four weather variables.

The actual (as opposed to potential) rate of evapotranspiration may match the potential rate if there is sufficient water in the soil to satisfy the demand from the atmosphere. If soil water content is low, however, the actual rate will fall below the potential rate (Figure 3.4).

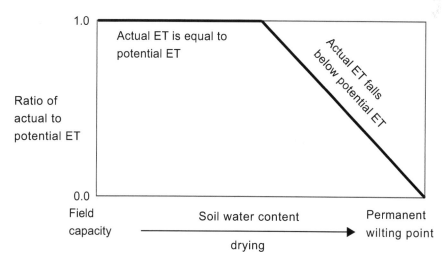

Figure 3.4 Potential and actual evapotranspiration rates

In contrast to rainfall, the spatial variation of potential evapotranspiration rates is relatively low, except in locations where climate varies over short distances (for example in areas of high relief).

Runoff and recharge

From a water resource perspective, runoff and recharge represent the amounts of water left over after evapotranspiration demands have been satisfied. In general (but with some exceptions), runoff and recharge occur when the soil is wet. The flow in streams and rivers consists of water that has run off the land rapidly during and shortly after rainfall events, when the ability of the soil to receive water in the form of infiltration is reduced relative to rainfall intensity; together with the natural discharge from aquifers adjoining watercourses (so-called 'baseflow'). It is groundwater discharge that keeps rivers flowing in dry seasons, so the linkages between surface water and groundwater are important. The '3R' approach, in which measures are taken to encourage recharge, so retaining water in the soil and groundwater system, and thereby allowing re-use of rainwater within the river basin, recognizes this principle (van Steenbergen and Tuinhof, 2010).

Recharge occurs both from rainfall (direct recharge, as shown in Figure 3.1) and from wetlands and watercourses (indirect recharge) following runoff (Figure 3.5). The proportions of these two processes vary depending on soil properties and land use, and the relative elevations of watercourses and aquifer water tables (the last of which may vary between wet and dry seasons).

From the point of view of maintaining water supplies, groundwater recharge is a key component of the water cycle, since it helps to determine how much water can be sustainably withdrawn from wells and boreholes. In areas where rural population densities are low, and water consumption correspondingly small, rural water supply generally does not pose unsustainable demands on groundwater resources. However, in dry regions, in locations where rural populations are steadily rising, and as demands for groundwater increase, it should not be taken for granted that water resources will remain sufficient.

In some locations (especially in dry climates where wide-scale irrigation is practised) the greatest threat to rural drinking water supplies does not come

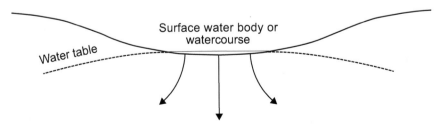

Figure 3.5 Indirect recharge from a watercourse to an adjacent aquifer

GROUNDWATER RESOURCES 37

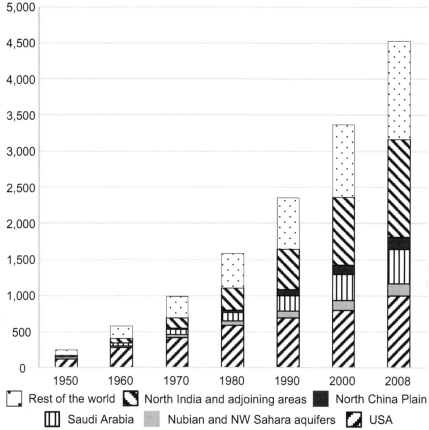

Figure 3.6 Estimated cumulative groundwater depletion by region since 1900, km³
Source: Data from Konikow, 2011, re-plotted

from its own abstractions, but rather from those of other consumers – especially those abstracting water for use in irrigation. Groundwater reserves in many parts of the world have been severely impacted by agricultural abstractions of groundwater (Figure 3.6), to the detriment of local rural domestic water supply (Foster et al., 2013; IAH, 2015).

'Green' and 'blue' water

The ideas of 'green' water and 'blue' water were first introduced by Falkenmark (1995) to distinguish between the water that is available in rivers and aquifers (blue water) and the water in the soil (green water), which is accessible to plants but not available for abstraction by humans. The distinction means that for rural drinking/domestic water supply, directly captured rainfall or abstracted surface water or groundwater are the only options.

In agriculture, however, measures to encourage infiltration of rainfall into the soil, so increasing the availability of green water (and at the same time enhancing groundwater recharge), are generally to be encouraged, especially in dry regions experiencing increasingly unpredictable rainfall patterns.

Consumptive and non-consumptive uses of water

Water that is evaporated from the land surface or open water bodies, together with that which is transpired by vegetation, is generally considered to be subsequently unavailable for local use (even though the water vapour involved is still involved in the water cycle). Water that follows this pathway, mainly in agricultural settings, is described as a consumptive use.

In contrast, some of the water that is 'consumed' in domestic and industrial settings is still available afterwards as liquid water, and is potentially available for re-use. In most cases this non-consumptive water has been contaminated, for example in the conversion of drinking water into urine, or handwashing or laundry water into so-called grey water. If it returns to a surface water body or to groundwater, it may be re-used, preferably after treatment.

Groundwater

Groundwater storage and movement

Groundwater is stored in, and moves slowly through, bodies of soil and rock beneath the earth's surface. Between the surface and the water table, the material is unsaturated, but below the water table in the aquifer proper, the rock is saturated. Water that directly recharges the aquifer passes through the unsaturated zone to reach it (Figure 3.1).

The term 'soil' is generally reserved for the top metre or so in which most roots of vegetation (with the exception of deep-rooting natural vegetation) are concentrated. Below this, there is often a weathered layer of rock, sometimes many metres or tens of metres thick, followed in turn by fresh rock at depth. The water table may lie within the weathered zone, or deeper.

Groundwater in rocks occupies the pore spaces between the solid components. The total amount of pore space (usually expressed as a proportion or percentage by volume) is its porosity. A more useful term, however, is the drainable porosity (also known as the specific yield) – the proportion of the rock volume that can drain readily by gravity. This idea recognizes that some water remains trapped in rock pores (especially the smaller pores) even after the water table has been lowered.

Even more important though than the drainable porosity is the storage coefficient (or storativity) of aquifers. Here the main distinction is between unconfined and confined aquifers. Figure 3.7 illustrates the difference between these two types of aquifer. An unconfined aquifer is bounded at the top by a water table, which is exposed to atmospheric pressure (as also in Figure 3.1). A confined aquifer holds water that is under pressure (like water in a pressurised pipe) such that, given the chance, the water

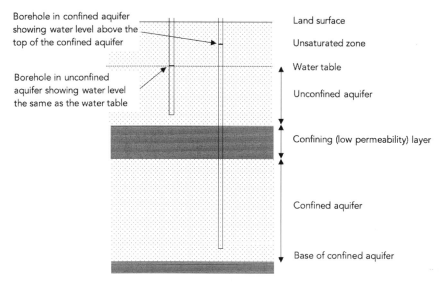

Figure 3.7 Key features of unconfined and confined aquifers

would rise above the top of the aquifer. One way to achieve this is to drill a borehole into the aquifer.

The storativity of an aquifer is the amount of water that is removed (per square metre of surface) as a consequence of a one-metre lowering of the water table or potentiometric surface. The potentiometric surface in a confined aquifer is the surface joining the water levels in wells or boreholes as shown in Figure 3.7. In an unconfined aquifer, the main mechanism involved when the water table is lowered by pumping is gravity drainage of pore space; therefore the storativity, S, is identical to the drainable porosity.

In the case of a confined aquifer, when the potentiometric surface is lowered by one metre, the pores of the aquifer remain full (the aquifer is saturated) before and after the pumping that has led to a lowering of the water level. This seems at first sight to be paradoxical – the aquifer is 'full' before and after the lowering of the water level, and yet some water nevertheless is pumped out. The answer is that two very small effects are happening, which also happen in unconfined aquifers, but there they are so small as to be insignificant alongside the draining of pore space. The two effects are, first, the elastic expansion of water as its pressure is reduced; and second, the elastic compression of the aquifer structure (its physical architecture) as it has to take a greater proportion of the weight of the overlying layers of rock. This compression may sometimes lead to significant land subsidence when groundwater is pumped.

The storativity of a confined aquifer is, unsurprisingly, much smaller (three to five orders of magnitude smaller) than that of an unconfined aquifer. The practical expression of this is that when water is pumped at the same rate from the two different types of aquifer, other things being

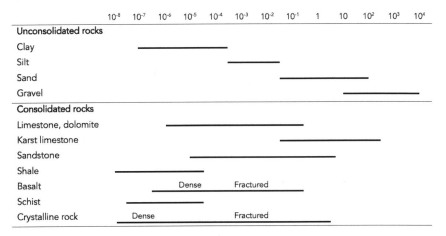

Figure 3.8 Hydraulic conductivity ranges of selected rock types, metres per day
Note: The extremely wide range of values – more than 10 orders of magnitude from the least permeable to the most permeable materials – is particularly significant
Source: Lewis et al., 2006

equal, a confined aquifer would experience a greater drawdown (reduction in water level due to pumping) extending over a larger area than in the case of an unconfined aquifer. It also responds much more quickly to changes in pumping rates.

The ease with which water can move through an aquifer depends on both the pore size and the degree of interconnectedness of those pores. How easily water can move through the aquifer is described by its hydraulic conductivity (Figure 3.8), often referred to loosely as the permeability. The hydraulic conductivity is an important determinant of borehole 'yield' (see below).

Intergranular or primary pores (the spaces between grains of rock in sedimentary and other rocks) range from sub-micron size in clays to several millimetres in sands and gravels; they are generally well connected. It is the coarser grained rocks (with larger pore sizes) that therefore form important aquifers in the case of sedimentary and other unconsolidated rocks. Choosing sites for wells or boreholes in such formations is generally straightforward from a hydrogeological point of view (Carter et al., 2014).

Pores in hard igneous and metamorphic rocks and those caused by dissolution in carbonate rocks (limestones and dolomites) have similar size ranges, but the degree of interconnection may be variable, in some cases meaning that different zones of the aquifer are essentially isolated from others. Useful hard rock aquifers are therefore those with very extensive and well connected fracture or dissolution patterns; otherwise only limited (but locally important) abstractions are possible. Siting of new wells or boreholes may be more complex in these geological formations (Carter et al., 2014).

How water moves in aquifers

Water moves through aquifers according to two very simple physical laws. The first is the conservation of mass – water cannot be created or destroyed; it all has to be accounted for. The second is Darcy's Law, the experimental finding that the flow rate in an aquifer is the product of (a) the hydraulic conductivity (Figure 3.8), (b) the cross-sectional area through which flow takes place, and (c) the hydraulic gradient. The hydraulic gradient is the slope of the potentiometric surface (which is the water table in an unconfined aquifer).

When these two physical laws are combined, they result in second order differential equations, which can appear very daunting to a non-mathematician despite the simplicity of their origins. Moreover, the equations so produced cannot be solved directly, and they require solution either by numerical techniques or via approximations. Each of these options has strengths and weaknesses. Numerical solutions (as applied in digital computer models of groundwater) have been described as 'approximate solutions to the exact equations', while the development of so-called analytical solutions (approximate equations) has been described as 'exact solutions of approximate equations'. The mathematics of groundwater flow, its applications in the analysis of pumping tests on wells and boreholes, and its application in groundwater modelling, lie beyond the scope of this book, but are thoroughly dealt with in a range of hydrogeological text books (see, for example, Freeze and Cherry, 1979; Fetter, 1994; Price, 1996; Macdonald et al., 2005; Anderson et al., 2015).

Natural flows
In the absence of pumping, groundwater moves in accordance with natural hydraulic gradients and its rate of movement is determined by the hydraulic conductivity of the aquifer material. As water percolates from the earth's surface to an underlying aquifer (by direct recharge), the hydraulic gradient may be high, but the hydraulic conductivity of the unsaturated soil and rock is low – unsaturated hydraulic conductivities are several orders of magnitude below those occurring in saturated materials (Figure 3.9). Accordingly, vertical rates of movement can be extremely low (sometimes much less than one metre per year), except where water follows so-called preferential flow paths such as cracks and root channels. These preferential pathways are very common, and even in soils and rocks that lack such physical features, water may move downward along wetter 'fingers' within otherwise drier soil (Jarvis et al., 2016). Consequently it is possible for contaminants near the surface to reach the water table more quickly than would be calculated by assuming a uniformly low (unsaturated) hydraulic conductivity for the materials overlying an aquifer.

Beneath the water table in the saturated zone, hydraulic gradients occurring in nature are generally low, except in areas of pronounced relief or near to areas of natural groundwater discharge. Consequently, groundwater flow velocities

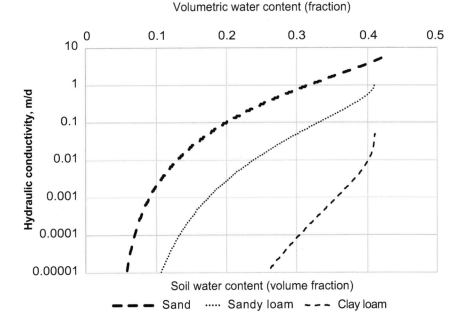

Figure 3.9 The relationship between hydraulic conductivity and water content
Source: Equations in van Genuchten and Pachepsky, 2011, and soil properties in Bonan, 2019

in nature are often low too – a few metres per day compared to velocities in streams and rivers, which may be up to a few metres per second.

One important consequence of the typically slow flows in aquifers is that when something changes (for example the rate of recharge or the rate of pumping), it may take a long time (decades or even centuries) for a new equilibrium – a balance between inflows and outflows – to be reached (Cuthbert, Gleeson et al., 2019). Changes taking place today may only be felt by future generations a hundred years or more from now.

Groundwater flows in response to pumping
When water is removed from a well or borehole, a cone of depression or drawdown curve is induced in the surrounding aquifer, the depth and extent of which depend on the pumping rate and duration, and the properties of the aquifer (Figure 3.10). In practical terms, there are limits to the drawdown that can be allowed, primarily depending on the thickness of the aquifer (Figure 3.11; see also Chapter 4).

The key aquifer property is the transmissivity, the product of the hydraulic conductivity (Figure 3.7) and the vertical saturated thickness of the aquifer. The value of transmissivity is determined by carrying out a pumping test on a newly constructed well or borehole. The greater the transmissivity, the higher the rate of abstraction that can be supported (Table 3.1).

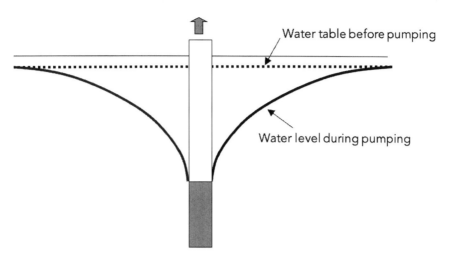

Figure 3.10 The cone of depression or drawdown curve around a well or borehole

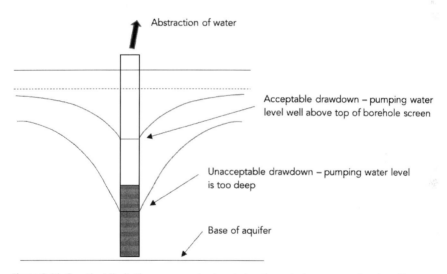

Figure 3.11 Practical limitation on groundwater abstraction rate in an unconfined aquifer

Table 3.1 Aquifer potentiality as a function of aquifer transmissivity

Transmissivity (m^2/d)	Aquifer potentiality	Comment
<5	Negligible	Nevertheless enough for a handpump abstracting up to 0.3 l/s
5–50	Weak	Transmissivity of 50–100 m^2/d or more is sufficient to
50–500	Moderate	support an abstraction rate of 5 l/s
>500	High	Sufficient for higher demands such as small town water supply

Source: de Wiest, 1965; comments by this author

Groundwater abstraction rates – 'yield'

Two factors limit the amounts of water that may be abstracted from wells or boreholes penetrating groundwater aquifers. The yield of each individual well or borehole is limited first by the hydraulics of groundwater flow towards it (Figure 3.11). The abstraction rates for individual wells or boreholes physically cannot exceed what the aquifer can transmit, as determined by the transmissivity, storativity, and other physical properties. If more water is needed than is possible from a single well or borehole (because drawdown would be too great), then multiple wells may be needed. However, the second constraint, the overall amount that can be taken without causing adverse impacts on the environment or on other users, may then come into play.

This second limitation is the rate at which water can be pumped without bringing about unacceptable environmental or social impacts (including impacts on wetlands, for example, as well as on other abstractions). There has been a long-running debate about what constitutes a 'safe' or 'sustainable' yield in this sense, and there is no simple or generalizable answer, since all abstractions have implications for other parts of the water cycle, and new equilibria in groundwater take many years or more to become established when conditions change (Zhou, 2009; Cuthbert, Gleeson et al., 2019). Under 'natural' conditions, inflows to the aquifer (in the form of recharge) are balanced by outflows. When water is pumped from the aquifer, discharge may change, but so too may the recharge. Eventually a new equilibrium may be achieved. The estimation of what constitutes a safe or sustainable yield from an aquifer is not straightforward, but the principle still applies that there is a finite limit on the aggregate rate of abstraction for multiple wells or boreholes pumping from the same aquifer.

In the Africa-wide study carried out by MacDonald et al. (2012), the authors found that well and borehole yields (in the sense of the actual rate of pumping from water supply boreholes) are almost everywhere sufficient to support handpumps (delivering 0.1–0.3 litres per second); yields greater than 5 litres per second are relatively uncommon (except in the deep sedimentary aquifers of north Africa); but 'the potential for intermediate borehole yields of 0.5–5 litres per second, which could be suitable for small scale household and community irrigation, or multiple use water supply systems, is much higher'. A more recent modelling study in Ghana (Bianchi et al., 2020) found results broadly consistent with these conclusions. Table 3.2 is an attempt to translate such yield statistics into their implications for rural and small town water supply.

Trends and change

Water resources, in common with many other aspects of the environment, are experiencing changes as a consequence of two main pressures: population growth and man-made climate change. In regard to water quantity, both these pressures directly affect aspects of the water balance. Population

Table 3.2 Well or borehole yield and use

Well or borehole yield (l/s)	Abstraction and use	Typical volume pumped (m³/d)	Number of people served (domestic supply)
0.1–0.3	Communal handpump for domestic use	3–8	200–500*
1.0–5.0	Petrol, diesel, or solar pump for rural or small town supply	30 150	300–600** 1500***
5.0	Pumping solely for irrigation	150	approx. 3 ha irrigated#

* Assuming per capita consumption of 15 litres per day
** Assuming per capita consumption of 50–100 litres per day
*** Assuming per capita consumption of 100 litres per day
\# Assuming crop water requirement of 5 mm per day
Note: Estimates assume eight hours of pumping per day

growth leads to changes in land use, including increased clearance of land for building construction and for extending and intensifying agriculture. Climate change is reflected in rising temperatures, increases or decreases in mean annual rainfall (depending on location), and intensification of rainfall, among other changes.

Population

While population growth in most regions of the world has either levelled off or is projected to do so by the middle of the 21st century, Africa continues to experience a sustained rate of growth (Figure 3.12). In general, rates of urban population growth are higher than those in rural areas, but rural population in Africa is still rising. More significantly for rural water supply, African rural populations are likely to become increasingly represented by less mobile people, who may be elderly, less well-educated, or suffering from ill health or disabilities that prevent them seeking opportunities in towns and cities. This may have implications for community management of rural water supply in the decades following the end of the SDG period (see Chapter 7).

In some countries, even rural population densities are starting to reach levels at which groundwater abstractions for domestic use may soon start to exceed sustainable limits. In rural areas of Malawi, for example, communities are heavily dependent on groundwater for their domestic water supply. The national renewable groundwater resources are estimated to be 2.5×10^9 m³ per year (UNEP, 2010). This translates to a depth-equivalent of about 21 mm per year, or about 14 times the likely depth-equivalent abstraction rate for rural water supply of around 1.5 mm per year (based on a population density of 200 persons per km² and a per capita consumption of 20 litres per day). On the face of it, this suggests that Malawi's rural handpumps should not be resource-limited in terms of their ability to support their populations.

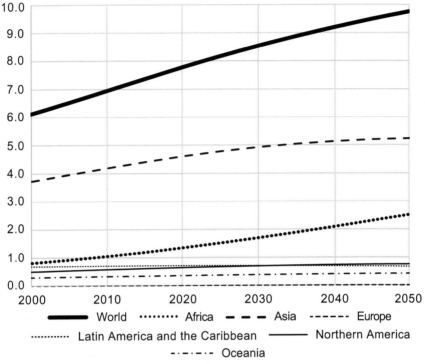

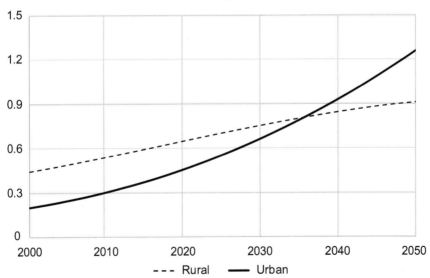

Figure 3.12 Population projections (billion people), 2000–50, by region and within sub-Saharan Africa
Source: UN DESA, 2018

However, this assumption has been challenged (Davies et al., 2013) on the grounds that recharge within the limited capture zones of boreholes may often be insufficient to support abstractions, especially as water demands grow over time and successions of drought years are experienced more frequently.

Climate change

A number of aspects of man-made climate change can be stated with high confidence (IPCC, 2020):

- During the industrial period land surface air temperatures have risen by nearly twice as much as global average temperatures.
- Climate change, including increases in the frequency and intensity of extremes (droughts and intense heavy rainfall), has already adversely impacted food security and terrestrial ecosystems in Asia, Africa, Europe, and South America.
- The future risks of droughts, water stress, heat-related events, and land degradation are likely to increase if greenhouse gas emissions continue unabated.
- Asia and Africa are projected to have the highest numbers of people vulnerable to increased desertification; coastal areas are vulnerable to sea level rise; in cyclone-prone areas lives and livelihoods will be increasingly at risk.

Changes in rainfall patterns (total amount, spatial and temporal distribution), combined with changes in other weather variables that affect evapotranspiration, make the prediction of changes in runoff and recharge both uncertain and location-specific. When combined with demographic changes and uncertainties about how rural populations will change their use of land in future, a 'cascade of uncertainty' (Mitchell and Hulme, 1999; Wilby and Dessai, 2009) follows in regard to estimates of water resources. This places a strong imperative on continuous and adaptive monitoring of water resources by national institutions. All water sector actors should supply relevant data to national databases, and such databases should be open and accessible to those having a legitimate interest in such information.

CHAPTER 4
Water supply boreholes

Abstract: *The creation of water supply boreholes requires competent siting, which takes account of both the hydrogeology and the needs of water users or system operators. The design of the borehole begins with an understanding of the use(s) to which water will be put and the quantities needed. The design must specify the depth, diameters (of drilling, of lining), and materials to be used in the construction of the borehole. Boreholes may be drilled by manual or mechanical techniques, with the selection of appropriate method being based on the ground conditions and factors such as cost, site access, and how critical the speed of construction is. Drilling and completion of the borehole must be undertaken by trained, experienced, and certified technicians whose practices and required qualifications are set out by professional drillers' associations and by governments. The construction of water supply boreholes that are fit for purpose must be supervised by competent and experienced hydrogeologists or engineers.*

Keywords: boreholes, siting, design, drilling, completion, development, test pumping, water quality sampling, supervision

> 'the lowest cost well is not always the most cost-effective, particularly if construction quality is compromised to save money'
> —Danert et al., 2010

Introduction

Thousands of pages of published textbooks and manuals set out the details of water supply borehole construction. A select list is included at the end of this chapter. It is not my intention here either to repeat or summarize what has been said more than adequately elsewhere. Rather my aim is to highlight the most critical aspects of the subject, so that common mistakes are not perpetuated. This chapter is not a treatise on groundwater engineering, but I do provide sufficient technical detail to help the reader understand the significance of the main subjects highlighted.

As far as this topic is concerned, much knowledge exists, and it is well documented. Two priorities exist: first, to ensure that all those involved in borehole construction acquire and assimilate that knowledge; and, second, that they apply it rigorously.

An important consequence of shoddy construction is that some completed and commissioned boreholes are simply not fit for purpose. Because of poor siting decisions, yields may be insufficient or water quality unacceptable.

Boreholes may dry up seasonally because they were not constructed to a sufficient depth. All these factors affect communities' or operators' motivations to use and maintain water points, and their jobs may become difficult or impossible.

Before going further, a brief note is needed on terminology and on the scope of this chapter. The term 'well' is used with two alternative meanings: either it refers specifically to a structure of large diameter (large enough to accommodate human well-diggers); or, in North American usage, it refers to any hand-dug, driven, or drilled excavation to access groundwater. The word 'borehole' refers more specifically to a slim, manually drilled or machine-drilled water point. The term 'tubewell', used in many Asian contexts, is synonymous with borehole.

This chapter is about those relatively small-diameter drilled or driven water points – boreholes or tubewells – which are the major sub-set of water supply wells and boreholes. The chapter does not generally deal with hand-dug wells, except where matters affecting the sustainability of services provided by boreholes are also true of those large diameter structures.

Overview – what is required?

In order that boreholes are indeed fit for purpose, a number of conditions need to be fulfilled. Boreholes may be equipped with handpumps, so acting as point sources, or when equipped with mechanised pumps they may supply piped systems that serve rural areas or small towns. In either case they must:

- be in the right place (from either the users' point of view in the case of water points, or from the engineering point of view in the case of piped systems);
- be able to access sufficient groundwater of acceptable quality;
- be designed correctly, to meet the water needs of the users;
- be properly constructed to well established standards;
- be cost-effective – not necessarily least-cost, but at a cost commensurate with good quality design and construction.

To achieve these outcomes, a number of processes need to be followed. First, specifications relating to the finished product and the processes needed to create it must be clear and unambiguous since these form an important part of construction contracts. Second, contract terms and conditions need to be drafted in ways that incentivise high quality work and deter short cuts. Third, competent contractors must be engaged to undertake construction. And, finally, supervision of construction needs to be carried out conscientiously and professionally.

The influential 'Rural water supply network code of practice for cost-effective boreholes' (Danert et al., 2010) sets out nine principles broadly covering the who, what, and how of the process for creating boreholes which are fit for purpose (Box 4.1).

> **Box 4.1 Nine principles for cost-effective boreholes**
>
> **Principle 1** Professional drilling enterprises and consultants – construction of drilled water wells and supervision is undertaken by professional and competent organizations that adhere to national standards and are regulated by the public sector.
>
> **Principle 2** Siting – appropriate siting practices are utilized.
>
> **Principle 3** Construction method – the construction method chosen for the borehole is the most economical, considering the design and available techniques in the country of construction. Drilling technology needs to match the borehole design.
>
> **Principle 4** Procurement – procurement procedures ensure that contracts are awarded to experienced and qualified consultants and drilling contractors.
>
> **Principle 5** Design and construction – the borehole design is cost-effective, designed to last for a lifespan of 20 to 50 years, and based on the minimum specification to provide a borehole that is fit for its intended purpose.
>
> **Principle 6** Contract management, supervision, and payment – adequate arrangements are in place to ensure proper contract management, supervision and timely payment of the drilling contractor.
>
> **Principle 7** Data and information – high quality hydrogeological and borehole construction data for each well is collected in a standard format and submitted to the relevant Government authority.
>
> **Principle 8** Database and record keeping – storage of hydrogeological data is undertaken by a central Government institution with records updated and information made freely available and used in preparing subsequent drilling specifications.
>
> **Principle 9** Monitoring – regular visits to completed boreholes are made to monitor their functionality in the medium as well as long term, with the findings published.
>
> *Source*: Danert et al., 2010

The following sections follow the process for creating wells and boreholes that will serve their communities for long periods. Such structures may last for 50 years or more under favourable circumstances – but only if the outlined requirements and principles are observed in full.

Choosing the site

The right place for a well or borehole is where two conditions are met: first, there is groundwater of sufficient quantity and quality for the intended uses and users; and second, in the case of a directly accessed water point, it must be conveniently located for its users. The process of siting and then creating a well or borehole must be seen therefore not simply as a technical matter, but one which lies at the interface of the social and the technical. If there is insufficient water, or the water is of unacceptable quality, the source will not be used; the same outcome follows if the water point is inconveniently located or at a place that deters people from using it.

From a hydrogeological viewpoint, the selection of a specific site starts with an understanding of the geology of the area and the relationship of

> **Box 4.2 Ten key factors in well or borehole site selection**
>
> 1. Sufficient yield for the intended purpose.
> 2. Sufficient renewable water resources for the intended purpose.
> 3. Appropriate water quality for the intended purpose.
> 4. Avoidance of potential sources of contamination.
> 5. Community preferences, women's needs, and land ownership.
> 6. Proximity to the point of use.
> 7. Access by construction and maintenance teams.
> 8. Avoidance of interference with other groundwater sources and uses.
> 9. Avoidance of interference with natural groundwater discharges.
> 10. A quantification of the risk of drilling a dry hole.
>
> *Source*: Carter et al., 2014

groundwater to that geology. In some geological formations groundwater can be found in good quantity and quality almost regardless of location (an exception being proximity to sources of potential contamination such as latrines or cemeteries). In others, good quality groundwater may be harder to find, being located in fractures, or zones of higher hydraulic conductivity relative to the surrounding rock. The range and sophistication of the scientific tools for siting consequently depend on the complexity of the hydrogeology. In simple well-understood formations, it may be enough to base the site selection on the geological or hydrogeological map. In more complex areas geophysical techniques (especially electrical resistivity and electromagnetic techniques), with or without the use of remote sensing, may be needed.

The Rural Water Supply Network's (RWSN) publication on siting of boreholes lists 10 key factors that need to be fulfilled in the selection of a good site (Box 4.2). With hindsight, the social aspects could have been emphasized more strongly, but these are elaborated further below.

Those who establish contracts for borehole siting with hydrogeologists need some awareness of the geology and hydrogeology of the areas where they are working, and the range of techniques available to hydrogeologists. Hydrogeologists often use geophysical equipment, sometimes together with satellite imagery; they interpret the data from their preparatory investigations and field surveys to produce recommended sites for drilling. A good hydrogeological survey is a highly cost-effective investment, and no drilling programme should be undertaken without one.

However, sometimes those siting reports are of limited real value, especially when the hydrogeologist lacks experience or integrity. Sometimes the use of geophysical techniques is unnecessary since groundwater is ubiquitous in the area in question. Sometimes conclusions that are drawn by the hydrogeologist are unjustified by the survey data. The programme manager or client needs to be able to distinguish between professional hydrogeological services and those which can too easily blind the client with (pseudo-)science. A useful guide for programme managers is the aforementioned RWSN publication (Carter et al., 2014).

Turning to the social or user aspects (points 5 and 6 in Box 4.2), two important matters stand out. First, the water point needs to be in the right place for the water users, avoiding locations such as cemeteries or sacred places, which may be unacceptable to the community. Second, it is important to ensure that access to the site will be possible both during its construction and in perpetuity. This may require that a written agreement is drawn up with the landowner, perhaps vesting title in the community or another entity.

It is easy for development professionals with a natural science or engineering background to be blind to the power struggles, hierarchies, divisions, and vested interests that exist within communities. Consulting with 'the community' over the location of a planned water point is far from straightforward. Who should be consulted? The leadership? The women? Those with disabilities, diseases, or infirmities, which may affect their access to water? The men? The wealthier individuals? The poor? The short answer is 'all of the above', and in a transparent and open manner. The details of how this can best be done are inherently country- and location-specific.

Design

The design of any engineered structure has to address two overarching questions, namely what is the structure meant to do (and avoid doing)? And how, precisely, will it achieve that?

A well or borehole needs to fulfil the following criteria: it must deliver the required discharge with an acceptable (i.e. not too great) drawdown, while providing structural stability, passing water (but not silt or sand), and protecting groundwater quality. To do so, it needs to:

- be drilled to an adequate depth;
- be of sufficient diameter to accommodate the anticipated pump;
- be lined as necessary with plain casing pipes and perforated well screens;
- be completed with suitable permeable backfill around the well screen;
- undergo an appropriate period of agitation or 'well development' to remove fine material and help restore damage caused by drilling;
- be finished with an adequate depth of impermeable sanitary seal around the upper part of the casing, and a suitable platform or apron at the surface.

Figure 4.1 shows how these features are reflected in a typical design for a water supply borehole in an unconsolidated aquifer. The fact that the geological formation is unconsolidated means that the borehole needs to be lined over its entire depth in order to prevent collapse. In the case of a hard rock aquifer, the lower part of the borehole is sometimes left open (unlined). It is important in such cases to be confident that subsequent ingress of mobile material (silt and sand) is very unlikely, and that collapse of the unlined interval will not occur. Some countries take a cautious view on this matter, requiring lining over the full depth of the borehole. The lining material

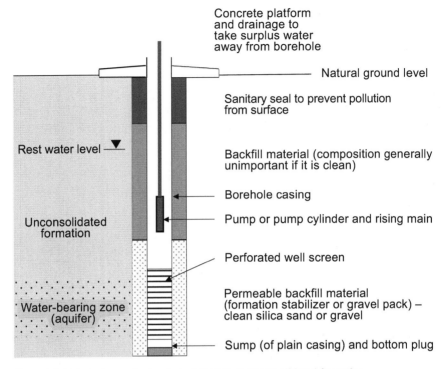

Figure 4.1 Main features of a borehole drilled in an unconsolidated formation

nowadays is commonly PVC, the well screen being made from PVC pipe having machine-cut slots, the size of which must be chosen by the hydrogeologist or engineer carrying out the borehole design.

Design procedure

A preliminary design needs to be undertaken in order to draw up specifications for inclusion in a contract with a driller. It is essential that the preliminary design is seen as exactly that. When the borehole is actually drilled, it is common to encounter unexpected conditions. The geology may be different to what was anticipated. Water strikes and the rest water level may be shallower or deeper than expected. The most permeable water bearing zones may be located at depths that are different to expectation. All this is entirely normal, and the design must be modified, on site, according to what is actually found.

A common mistake in borehole construction is that a standard design is included in the driller's contract, and the borehole is completed accordingly, without considering the actual geology encountered. This can result in productive layers being cased off, and well screens being placed adjacent to unproductive layers; it is one cause of poorly performing or low-yielding boreholes.

WATER SUPPLY BOREHOLES

The logic of borehole design is as follows (Figure 4.2). First, the geological sequence of the site should be described, together with the expected aquifer properties (especially the transmissivity and/or specific capacity – see Box 4.3) and rest water level. This information is gathered from geological and hydrogeological maps and reports, and from the logs of existing boreholes near to the location in question.

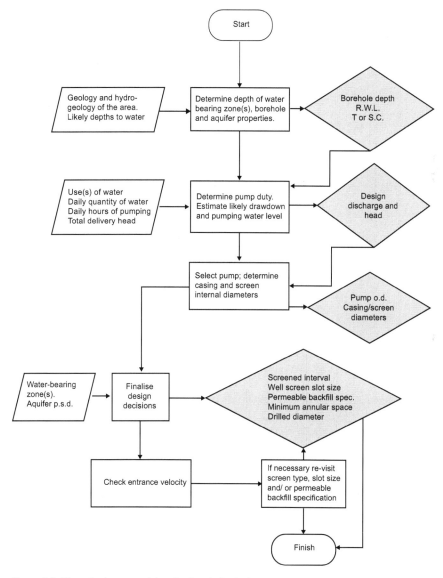

Figure 4.2 Flow chart summarizing the borehole design process
Note: R.W.L. rest water level; T transmissivity; S.C. specific capacity; o.d. outer diameter; p.s.d. particle size distribution

> **Box 4.3 Important aquifer properties**
>
> **Transmissivity** is a measure of the ability of an aquifer to transmit water to the borehole. It is the product of the hydraulic conductivity of the aquifer material (usually in m/d) and the aquifer's vertical thickness (m). Consequently, its units are m^2/d. Values in the tens or hundreds of m^2/d can provide good flows to boreholes, but lower values mean that significant drawdown is incurred. The drawdown as a result of pumping is inversely proportional to the transmissivity and directly proportional to the discharge.
>
> A very useful borehole property is its **specific capacity**. This is the discharge of a borehole (usually in m^3/d) divided by its drawdown (m). The borehole should have been pumped steadily, ideally for 24 hours or more, so that both the discharge and the drawdown have stabilized. If the average specific capacity for boreholes in an area is known (from previous drilling activities), it is very easy to infer the likely drawdown in a new borehole pumping at a given rate (see Box 4.4).

> **Box 4.4 Estimating drawdown and pumping water level during the design process**
>
> If there has been significant drilling activity in the area, and specific capacities of existing wells can be determined from their records, this provides a very practical way of estimating drawdown in a new borehole. The drawdown in the new borehole is its design discharge divided by the average specific capacity – being sure to use (and if necessary convert to) consistent units.
>
> In the absence of specific capacity data, the simplest equation relating the expected drawdown (s_w) to the borehole pumping rate (Q) and the aquifer transmissivity (T) is Logan's formula, given by $T = 1.22Q/s_w$. Rearranging the equation to make s_w the subject allows it to be estimated from a known or assumed value of transmissivity: $s_w = 1.22Q/T$. If the transmissivity of the aquifer is not known, but assumptions can be made about the hydraulic conductivity (K) of its material (see Chapter 3) and its likely saturated thickness (b), then the relationship $T = K.b$ can be used to fill that gap in knowledge.
>
> The expected pumping water level in the borehole is readily calculated as the rest water level minus the estimated drawdown.
>
> Note: There are many more sophisticated ways to estimate pumping drawdown; please refer to the texts listed at the end of the chapter for an overview of these

Next, the required discharge is determined. This depends on the purpose of water use (domestic use only or, for example, for livestock and small-scale irrigation too) and the daily quantity required, together with a realistic estimate of the number of hours of pumping per day. The required discharge must be informed by local experience about the typical rates that boreholes in the area can support. The design discharge, together with the anticipated aquifer properties, allows the estimation of the drawdown and pumping water level. Box 4.4 provides a simple means of estimating drawdown and pumping water level.

Third, a suitable pump should be selected, based on the design discharge and required total delivery head (see Chapter 5). The reason for selecting the pump at this early stage in the design is that we need to know the physical dimensions of that pump (in particular its external diameter). The pump diameter determines the internal diameter of the borehole casing, since the pump will have to be accommodated inside it (Figure 4.1). Table 4.1 gives some

Table 4.1 Design discharge, pump outer diameter (o.d.), and casing internal diameter (i.d.)

Pump type and discharge	Typical pump o.d.	Required minimum casing i.d.
Handpump, 1–2 m³/h (up to 0.5 l/s)	Up to 75 mm	100 mm
Submersible electric pumps		
1–3 m³/h (up to 1 l/s)	75 mm	100 mm
3–10 m³/h (up to 3 l/s)	100 mm	125–150 mm
10–50 m³/h (up to 14 l/s)	150 mm	175–200 mm
50–150 m³/h (up to 42 l/s)	200 mm	250–300 mm

Source: Data from ACF, 2005, and Davis and Lambert, 2002

general guidance on pump sizes in relation to design discharge. It is important always to check the details of the pump dimensions with the manufacturer or supplier, however, since the largest outer diameter may be at the cable guard, in the case of submersible pumps.

Fourth, the casing diameter is selected by allowing its internal diameter to be 25–50 mm greater than the greatest external diameter of the pump (or more, if borehole straightness and verticality are likely to be compromised). The well screen diameter is commonly the same as that of the casing, but in some cases it can be smaller in order to save money. There is a common misconception that the diameter of the borehole or of the well screen determines the borehole 'yield'. This is only true inasmuch as the casing diameter limits the size of pump that can be used. But the diameter of the borehole has very little influence on its ability to receive water from the surrounding aquifer; this is controlled by the hydraulic properties of the aquifer (Chapter 3) more than by the diameter of the borehole.

These first four steps can – and should – be undertaken before the start of drilling. The fifth step is to examine the drilling samples and determine the particle size distribution of the (most coarse grained) aquifer material where the well screen will be positioned. Five decisions arise from this:

1. The depth interval to be screened. This should avoid zones of small particle sizes that could later seep into the borehole; the intention is to screen coarse, permeable water-bearing layers.
2. The slot or perforation size of the well screen; this needs to be small enough to prevent ingress of fine material from the aquifer or permeable backfill.
3. The grain size specification of the permeable backfill to be used; this backfill may be simply a sand or gravel that is of similar or larger grain size to the aquifer (known as a formation stabilizer) or a true gravel pack, which is specifically designed to prevent passage of fine particles from a fine textured and uniform grain-sized aquifer; in the latter case the particle size distribution curve is usually designed by multiplying that of the aquifer by a factor between four and six.
4. The minimum thickness of the annular space around the well screen where the permeable backfill will be placed; this should be at least

50 mm in the case of a simple formation stabiliser, or 75–100 mm in the case of a true gravel pack.
5. The diameter for drilling is then the outer diameter of the casing plus two times the annular thickness; in the case of handpump boreholes, all such pumps fit inside 100 mm internal diameter casing – this means that a drilled diameter of 200 mm is needed; it is common however to drill a little smaller, at 150 mm.

Preliminary decisions taken in the design process enable required quantities of materials (such as casings, well screens, and backfill materials) and drilling consumables (including drill fluids, fuel, oil, and grease) to be purchased and taken to site at the time of drilling. It is important to allow for contingencies, however, in case more than anticipated of any of these supplies are needed before drilling actually takes place. It is essential to make intelligent decisions and modifications to pre-drilling expectations, and these decisions are ultimately the responsibility of the supervisor.

The final step in the design is to check that the velocity of flow, as water enters the borehole, is not excessive. High entrance velocities lead to high drawdowns (and hence high energy costs in pumping), and they can contribute to a number of other undesirable effects such as corrosion, encrustation, and biofilm build-up. The entrance velocity is easily calculated by dividing the design discharge (in m^3/d) by the open area of the well screen (in m^2). The open area of the well screen depends on its length, diameter, and the proportion of its open surface area (this value can vary from <5 per cent to around 50 per cent depending on its material and construction). The calculation gives a velocity in m per day, which is usually converted to m/s. Misstear et al. (2017) review the extensive discussions regarding entrance velocity that have been conducted in the groundwater literature, concluding that the entrance velocity should be kept below 0.05 m/s (or <0.03 m/s if corrosion, encrustation, or biofouling are deemed likely). Note, however, that the authors assume that half the available open area is likely to become clogged up over time; if the manufacturer's open area figures are used in the calculation, this has the effect of reducing the maximum permissible entrance velocity to half of the values just given, namely 0.025 m/s and 0.015 m/s.

If the calculated entrance velocity is found to exceed one of these limits, then the design has to be reviewed to see if the screen length, screen slot size, or screen open area percentage can be increased. If not, the discharge may have to be reduced.

The outcome of the design process is a set of design decisions, listed in the diamond-shaped boxes in Figure 4.2. These should all be included in the specification of the borehole, which will form part of the contract with the driller.

Drilling

Borehole drilling can be carried out by a range of manual and mechanical processes. Regardless of the technique used, drilling has to accomplish three things. It must: break and penetrate the ground; remove the spoil or cuttings from the hole; and prevent the hole from collapsing. The different drilling methods approach these tasks in a variety of ways; these are illustrated in Figure 4.2, which also shows some of the equipment commonly used. The figure is not exhaustive.

Manual techniques (augering, manual percussion, sludging, and jetting) are typically used to depth of about 25–30 m, but exceptional depths are possible (>100 m), especially with sludging. Because large amounts of energy are needed to break hard rock, these methods are generally less suitable in such geology. Recent publications by UNICEF (2010) and RWSN (Danert, 2015) provide extensive detail on the potential and reach of manual drilling.

Mechanical drilling (cable percussion, mud rotary, and down-the-hole (DTH) hammer drilling) extends the possible diameters and depths, and the speed with which drilling can take place. It is common for a water supply borehole to be drilled in little more than a day, even in hard rock.

Whichever drilling technique is used, the most important consideration is the competence of the drilling contractor. Consequently, a great deal of effort has been put into training of drillers and the establishment of professional drillers' associations. RWSN and UNICEF have been particularly active in such efforts (Danert et al., 2020).

Borehole drilling can be carried out by private sector drilling contractors, NGOs, and faith-based organizations, or by state-run enterprises. Direct drilling by the state is becoming less common, but the first two often co-exist. Where they do, private sector contractors often complain about the not-for-profit sector undercutting them in project tenders. There is truth in this accusation, and little doubt that the perpetuation of drilling by non-profit organizations can hinder the development of a vibrant private sector.

Completion

Following drilling and the removal of the drill rods and tools, a number of activities need to take place (refer to Figure 4.1):

1. The permanent 'string' of borehole lining materials needs to be installed. This consists, in order of entry into the borehole, of the end cap and sump, the well screen, and the casing.
2. The permeable backfill is placed, taking care to ensure that it reaches the bottom of the hole and extends upwards to a few metres above the top of the well screen (to allow for settlement).
3. Any additional backfill is placed, to a level about 5 m below ground;
4. The sanitary seal, consisting of a neat cement slurry, bentonite (clay), or a cement-bentonite mix is placed up to the surface.

Probably the two aspects of completion that are carried out least well in practice are the selection of material for the permeable backfill, and the placement of an effective sanitary seal. The permeable backfill should be clean, well-rounded silica sand or gravel. Any material containing large amounts of fine particles, or any material that will break down to form such particles, is not suitable.

60 RURAL COMMUNITY WATER SUPPLY

 A screw is turned into the ground by successively adding more drill pipes to the top of the string. After a few turns, the entire drill string needs to be removed for emptying. Not suitable in gravel, rock, or collapsing ground. Sometimes used to reach the water table before changing to another manual method, but rarely used for the entire depth of drilling.

(a) Manual augering

 A set of heavy tools suspended by a rope (manual percussion) or a cable (machine drilling) is successively raised and dropped to break ground and, with some tools, grab spoil. The tools need to be lifted for emptying. Suitable in many types of geology, but slow in hard rock.

(b) Percussion

 Water (sometimes mixed to a viscous fluid by use of drilling mud) is pumped down a drilling pipe. It emerges at the tip and washes back up the hole, carrying cuttings to the surface. If the ground is prone to collapse, temporary casing can be inserted as drilling progresses. Suitable in unconsolidated formations.

(c) Jetting

WATER SUPPLY BOREHOLES 61

A steel pipe is raised and lowered in the borehole using a lever, while another crew member alternately covers and opens the top of the pipe with the palm of their hand. The hole is kept water-filled. As drilling progresses, a slurry of water and drill cuttings is lifted in the drill pipe. A common technique in unconsolidated formations in Asia.

(d) Sludging

Viscous drill mud is pumped down the drill pipe, emerging at the bit and washing cuttings up the hole. As with the manual techniques of jetting and sludging, mechanized (direct circulation) mud rotary drilling relies on the hydrostatic pressure of the fluid in the borehole to prevent collapse.

(e) Mud rotary

A compressed air-powered hammer strikes rapidly on the bottom of the hole, while simultaneously being slowly rotated. The escape of the compressed air from ports in the drill bit carries drill cuttings as small fragments or dust to the top of the hole. The main component of a drilling rig for DTH drilling is an air compressor. The method is especially suitable in hard rock.

(f) Down-the-hole hammer

Figure 4.3 Main water borehole drilling methods

62 RURAL COMMUNITY WATER SUPPLY

Crushed aggregate, road stone, or laterite, for example, are unacceptable. The sanitary seal is crucial for preventing ingress of surface pollutants. Without an adequate impermeable seal, contamination by pathogens of human or animal faecal origin is likely.

A further note on specifications

The design decisions highlighted in Figure 4.2 give rise to a number of components of the borehole specification. Further content of that specification relates to the completed structure. Regardless of the drilling method that is used to make the borehole, it must satisfy the following criteria. It must be straight (that is, not banana-shaped) and vertical (not inclined) (Figure 4.4). It must have undergone a sufficient period of 'well development' or agitation, the purpose of which is to remove fine particles introduced by the drilling process (including in any drilling fluids used), and leave behind a high-permeability zone around the well screen (Figure 4.5). It must be tested for yield, by undertaking a pumping test of suitable format and duration (Box 4.5). And, finally, it must be sampled for water quality analysis, with testing taking place on site for certain parameters, and others in a quality-assured laboratory (Box 4.6).

Straightness and verticality

Maintaining straightness and verticality during drilling is important, so that the placement of well screens and casings is achieved easily, and so that the pump does not rest on one side of the completed hole, so leading to wear and possible damage to the borehole lining and pump. The skill of the driller is crucial here.

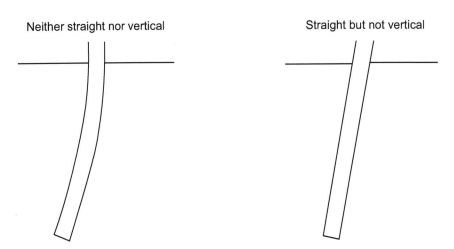

Figure 4.4 Borehole straightness (alignment) and verticality (plumbness)
Note: Both sketches are exaggerated

WATER SUPPLY BOREHOLES   63

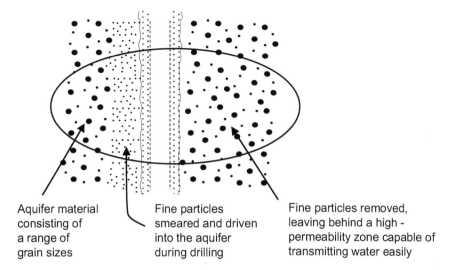

Aquifer material consisting of a range of grain sizes

Fine particles smeared and driven into the aquifer during drilling

Fine particles removed, leaving behind a high-permeability zone capable of transmitting water easily

Figure 4.5 The desired outcome of well development

Box 4.5 Test pumping

Ideally, two types of pumping test are conducted. The first is a step-drawdown test, in which the borehole is pumped at three or more different rates while measuring the drawdown. This kind of test allows an estimation of the borehole efficiency – a measure of how readily water enters the borehole. This can be useful as a baseline measure of performance, and this type of test may be repeated some years later if clogging, encrustation, or biofilm build-up is suspected.

The second, longer, test (a constant discharge test) is carried out at a constant pumping rate (typically about 10% higher than the design discharge). As with the step test, drawdown is measured regularly during the test, and discharge is also checked from time to time. When the pump is switched off, the water level recovers towards its original rest water level, and the progress of this recovery should also be monitored. The constant discharge/recovery test enables calculation of the aquifer transmissivity.

Test pumping practice is set out comprehensively by Kruseman and de Ridder (1994).

Well development

Well development has been described as 'the act of repairing damage to the borehole caused by the drilling process and removing fine-grained materials or drilling fluids, or both, from formation materials so that natural hydraulic conditions are restored and well yields enhanced' (ASTM, 2015). The process of well development – achieved by agitating, surging, over-pumping, or other techniques – helps to remove fine particles, which have come from different parts of the formation, and from drill mud in the process of drilling. Shortcuts are frequently taken by drillers in this area, and so special attention should be paid by drilling supervisors to the adequacy of well development.

> **Box 4.6 Water quality sampling**
>
> The main water quality parameters of interest fall into three categories: physicochemical parameters including pH (the main measure of acidity), Eh (the measure of oxidising/reducing potential), temperature, dissolved oxygen, and electrical conductivity (indicative of total dissolved solids); chemical species (inorganic and organic); and microbiological indicators.
>
> The manner in which the sample is obtained, stored, preserved, and transported can have major impacts on its usefulness and representativeness. Samples that are stored in contaminated bottles, not kept chemically preserved or at a sufficiently low temperature, or that fail to reach the laboratory quickly may give spurious results. This is especially true for the microbiological indicators, which are central to drinking water quality assessment.
>
> Some parameters such as pH and dissolved oxygen must be measured on-site in order to be representative of the groundwater. Furthermore, arrangements should be made to test them before the groundwater has become exposed to the air.
>
> Laboratories do not always operate to the highest standards, and so the inclusion of a few samples for which the analysis is already known (for example pure water) can be a useful way of exposing errors arising from contamination or poor laboratory practice.

Figure 4.5 shows in schematic form the condition of the aquifer adjacent to the well screen just after completion of construction (on the left-hand side), and the desired result of development (on the right-hand side). The fine particles surrounding the well screen and permeable backfill have been drawn into the borehole and pumped away to leave a high permeability zone around the well screen.

Test pumping

Once development is complete, the borehole must be test pumped. This is an important stage, since it permits the final determination of the allowable discharge of the borehole, and it enables important properties of the aquifer (notably transmissivity) and the borehole itself (specific capacity and hydraulic efficiency) to be calculated (Box 4.5).

Water quality sampling

Water sampling, testing, and laboratory analysis is a huge topic, the detail of which is beyond the scope of this book. A lengthy chapter in Misstear et al. (2017, ch. 8) explains in depth the why and how of water sampling and analysis, and Box 4.6 highlights some of the key issues for rural water supply in low- and middle-income countries.

Supervision of construction

The preceding sections of this chapter have outlined how borehole construction should ideally be done, and some of the ways in which a satisfactory outcome can be achieved. It is the role of the supervisor, working closely (but not in collusion) with the driller, to ensure that this result is actually achieved.

There has been an unfortunate failure to resource the training of hydrogeologists, thereby limiting the pool of potential supervisors. One consequence of this has been the tendency to frame drilling contracts as no-water-no-pay arrangements. It has been assumed that by placing all the risk on the driller (rather than sharing the risk between the client and the contractor), close supervision becomes less necessary – a convenient but nevertheless shaky assumption. If supervision was lacking, and the borehole is in fact dry, the driller can nevertheless make many incorrect claims about the presence of water, the depth drilled, the materials installed in the borehole, and how post-construction activities were conducted.

It is absurd to establish a contract with a driller and blindly trust them to undertake construction conscientiously and without cutting corners; if an individual were having a house built, he or she would oversee the construction carefully to ensure that the outcome is as desired. A borehole is no different in this regard.

Drilling for water is inherently risky, in some places riskier than others, but nowhere is success guaranteed. The hydrogeologist can do the best job possible in terms of the selection or confirmation of the chosen site, but the sub-surface is unknown in detail until the borehole is actually drilled. If there is no water (its quantity is insufficient or its quality is unsuitable) I believe the driller should be paid for the considerable cost of mobilization and demobilization of equipment and drilling. But this can only happen if it is known reliably what the driller has done. Consequently, the sector needs more well-qualified and conscientious supervisors.

The RWSN publication on supervision of water well drilling (Adekile, 2014b) sets out nine key responsibilities of the supervisor, corresponding to steps in the creation of a new borehole and equivalent responsibilities of the driller (Box 4.7). In the early years of the 21st century, a number of new training initiatives have been established to provide site-based and online short courses for drilling supervisors. This is to be welcomed, but until the number of hydrogeologists in lower-income countries increases significantly, with inclusion of this important topic in their course curricula, it is likely that some drilling will continue to result in poorly constructed boreholes. Moreover, the actual completion details are likely to be incorrectly represented in completion reports.

The way in which drilling services are procured, the forms of contracts between the client and the driller, and the subject of costing and pricing are addressed in two further publications from RWSN, namely Adekile (2014a) and Danert et al. (2014).

Data

Two of the principles for cost-effective boreholes (Box 4.1) emphasize the importance of data and information from drilling operations (siting reports, drill logs, borehole completion details, test pumping records, and water

> **Box 4.7 The supervisor's responsibilities**
>
> - Inspect equipment and interview personnel.
> - Check equipment; provide guidance on siting the borehole; approve siting report.
> - Together with the client, thoroughly discuss the design, materials, and procedures for each step of the contract.
> - Liaise with the community; approve drilling equipment and material; guide driller to site.
> - Monitor drilling; advise depth to stop drilling; log the borehole.
> - Instruct screening and casing depths; ensure gravel pack and sanitary seal properly placed.
> - Ensure water is clean; proper disinfection; supervise pumping test; ensure samples are taken and platform installed.
> - Ensure the site is restored to its former state.
> - Hand over borehole to community.
> - Report to client.
>
> *Source*: Adekile, 2014b

quality data) being submitted to national authorities, with those authorities maintaining reliable and accessible records. The costs associated with failure to maintain good quality data and information are potentially very high.

Selected texts and other useful documents on borehole design and construction

Action contre le faim (ACF) (2005) *Water, Sanitation and Hygiene for Populations at Risk*, Paris: Hermann. <https://www.ircwash.org/sites/default/files/acf-2005-water.pdf> [accessed 12 September 2020].

Driscoll, F.G. (1986) *Groundwater and Wells*, 2nd edn, St Paul, MN: Johnson Screens.

MacDonald, A., Davies, J., Calow, R. and Chilton, J. (2005) *Developing Groundwater: A Guide for Rural Water Supply*, Rugby: ITDG Publishing. <http://www.developmentbookshelf.com/doi/book/10.3362/9781780441290> [accessed 12 September 2020].

Misstear, B., Banks, D. and Clark, L. (2017) *Water Wells and Boreholes*, 2nd edn, Chichester: Wiley Blackwell.

Rowles, R. (1995) *Drilling for Water: A Practical Manual*, 2nd edn, London: Ashgate Publishing.

A series of publications by the Rural Water Supply Network, some of which have been cited in this chapter, provides valuable and free documentation on aspects of the topic. These may be downloaded from: <https://www.rural-water-supply.net/en/resources/filter/2_9> [accessed 12 September 2020].

CHAPTER 5
Water lifting from wells and boreholes: handpumps

Abstract: *This chapter describes six low-lift (typically up to 35 m) handpumps suitable for individual households or very small communities; it also describes a further seven deep well (typically up to 45–50 m) handpumps. Out of the 13 highlighted handpumps, the designs for eight are in the public domain; the remaining five are commercially owned. All have submerged cylinders. International experience with handpumps over the last 40 years or more has raised important issues around (a) the viability of so-called village-level operation and (management of) maintenance or VLOM(M), more commonly referred to as community-based maintenance; (b) the wisdom of placing pump designs in the public domain; (c) the value of national-level standardization policies; and (d) quality control and quality assurance of replacement parts. Alongside these factors must be considered the issues of corrosion and functionality. Although some measures have been implemented to address issues of corrosion, the notion of functionality as a useful metric has been criticized. Detractors instead propose that the identification of what needs to be in place to achieve sustainable services from community-managed handpumps would be more productive. It is evident that handpumps will continue to be necessary for the foreseeable future, thus the importance of having an independent handpump 'champion' to maintain standards and documentation is highlighted.*

Keywords: handpumps, low-lift pumps, direct action pumps, deep well pumps, VLOM(M), public domain, standardization, quality control, quality assurance, corrosion, functionality

> 'The handpump option' – a retrospective from the 2020s
> —Arlosoroff et al., 1987

Introduction

While water may be lifted from large diameter hand-dug wells by ropes attached to any of a wide range of improvised vessels made from calabashes, leather, or sawn-off jerry cans, the small diameter of drilled boreholes requires correspondingly slim pumps. Pumps designed for lifting water from boreholes can also be used in hand-dug wells, thus permitting such wells to be covered and thereby protected from ingress of contaminants. Consequently, this chapter focuses on manually operated pumps suitable for boreholes.

In common speech, a 'pump' often comprises both the element that moves water by hydraulic action (the pump in the strict sense), together with the component, often driven by a non-hydraulic source of energy, which conveys some form of motive power to the pump via a prime mover. For example, a deep well handpump consists of a pumping element below water (the cylinder-plunger-footvalve assembly) together with the means of conveying human mechanical power to that cylinder (the handle, fulcrum, and pump rods). It is useful for basic understanding to distinguish pumps in the strict sense from prime movers with their energy sources; this chapter begins by doing so (Figure 5.1). However, from a practical point of view, it is the usual sense of pump-with-driver that is more useful; consequently, the main part of the chapter is about pumps in that broader meaning.

Borehole pumps (in the narrower sense) span a wide variety of pumping principles and designs. Many rural water pumps involve a plunger moving vertically within a cylinder (reciprocating pumps); but rotary pumps and diaphragm pumps are common too. Some interesting new technology based on well-established thermodynamic principles, but never before used in boreholes, is emerging in the early 2020s, which promises much reduced repair burdens for rural communities (see Chapter 6).

Borehole pumps may be driven using a variety of prime movers powered by renewable and non-renewable energy sources, each with its advantages

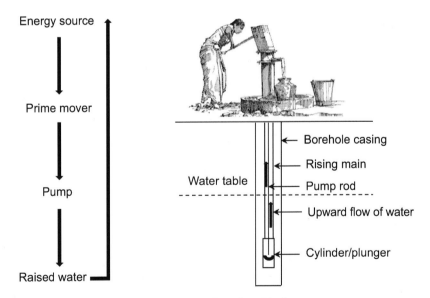

Figure 5.1 Water pump components, conceptually and practically
Note: The energy source is food; the prime mover is the pump operator; the pumping element is the submerged cylinder and plunger; power is conveyed to the plunger via the pump rods, with or without a lever
Source: Sketch courtesy of Rod Shaw (2015), WEDC, Loughborough University

and disadvantages. Each has its place, somewhere in the wide diversity of rural water supply contexts. Renewable energy sources include human, solar, and wind, while fossil fuels (usually diesel) may also be used if the circumstances require. Grid electricity can sometimes be used for water pumping. Various pumps exist that utilize the energy in a flowing or stored body of water for pumping. This chapter focuses on human energy sources for (rural) water pumping, while solar and other mechanized options are considered in Chapter 6.

The range of pumping options – pumps combined with energy sources – is immense, and an overview of the landscape and its main features is presented below. The story to date of rural water borehole pumping is long and it has some contentious features. The technology has to be considered within the wider context of its institutional, management, and financing implications, and its acceptability to water users. Consequently, although this is a chapter about technology, it has very human dimensions too. A recent summary of the last 50 years of handpumps in rural water supply has been developed by the Netherlands International Cooperation Collection (Holtslag, 2020).

Pump types

In their comprehensive handbook on water lifting, Fraenkel and Thake (2006) identify 35 types of pump, based on six fundamental operating principles. Despite that diversity, they point out that all water lifting devices can be characterized by their output at different heads (loosely, the lifts involved) and speeds (of rotation in the case of rotary pumps; stroke rate in the case of reciprocating pumps). In many cases there is an ideal combination of flow, head, and speed at which the pump is most efficient (in other words the output is maximized for a given power input). A number of pumps are of particular interest in the context of rural water supply (Table 5.1). Some further discussion of these follows here and in Chapter 6.

A selection of commonly used handpumps

In this section, I give a brief overview of a selection of handpumps in use at the time of writing. All but one of those which I describe (the exception being the Blair Institute Bucket Pump) have their cylinder or pumping element permanently located below water. Many more pumps have been omitted here for lack of space and limited familiarity of this author, including the Pitcher pump (Madagascar), the Bandung pump (Indonesia), and the Jibon pump (Vietnam). An excellent illustrated summary of all the pumps mentioned here (and more) is available on the Rural Water Supply Network (RWSN) website (n.d., b). It provides more detail than space allows here.

RURAL COMMUNITY WATER SUPPLY

Table 5.1 Main types of water pump used in rural water supply

Category and type	Examples	Comments
Direct lift		
Self-emptying bucket	The Blair Institute 'Bucket Pump'	Hardly used outside Zimbabwe
Displacement pumps		
Piston pumps	Most reciprocating handpumps	The dominant pumps used in rural water supply boreholes in lower-income countries. These are discussed in more detail below
Diaphragm pumps	The Vergnet Hydro pump	
Progressive cavity	The Mono pump	
Chain/rope and washer	The Rope pump	
Velocity pumps		
Rotodynamic pumps	Many models of centrifugal and mixed flow submersible pumps	Widely manufactured in China, Europe, India, and North America
Impulse pumps		
Hydraulic ram	The Impact Pump	See Chapter 6 'solar pumps'

Pumps for individual households or very small communities (low-lift)

Three pumps or groups of pumps stand out as particularly appropriate for small populations (one or a few households). All tend to be used at fairly shallow depth settings, notionally up to about 35 m. All lend themselves to local manufacture (rather than importation) and to straightforward repair processes that are often available in rural areas (with the exception of those very remote from communities with technicians).

The Bucket Pump was designed by the Blair Research Institute in Zimbabwe in 1983 (Morgan, 1990; WEDC, n.d.). It is the nearest thing to a rope and bucket, as would be used in a hand-dug well, but is designed for use in a slim borehole. The Bucket Pump (Figure 5.2a) consists of a cylindrical steel tube (the bucket) with a simple non-return valve at the bottom and a connection for a chain at the top. When the bucket is lowered into the well, it fills and is then raised by a windlass. It is placed over a simple water discharge unit, which pushes the footvalve up to release the water. The pump can be used at depths up to about 15 m, and it can deliver 0.10–0.15 litres of water per second. Given its simplicity, it is surprising that the Blair Bucket Pump has not spread significantly beyond Zimbabwe, where a few thousand installations have been made.

The Rope Pump (Figure 5.2b), however, has seen much wider application, especially in Central and South America and sub-Saharan Africa. A continuous loop of rope with flexible plastic or rubber disks or washers attached at intervals is turned, causing the washers to rise up a pipe, delivering water continuously. Modern (20th-century) improvements to an ancient pumping principle, together with its introduction to local commercial manufacturers in Nicaragua, and subsequent transfer to about 30 small manufacturers in six African countries, had led to a claimed 130,000 installations worldwide by

(a) The Blair Institute Bucket Pump (b) The Rope Pump

Figure 5.2 The Bucket Pump and the Rope Pump
Source: (a) Reproduced by permission of Peter Morgan, Zimbabwe

2016 (Haanen, 2016). The Rope Pump can be used at depths up to about 60 m with two handles, although it is mostly used at shallower settings with a single handle up to about 35 m. It has an output of 0.6 litres per second at 10 m depth, with that output halving at 20 m depth. At greater depths, the output is correspondingly reduced.

Improved Rope Pumps are made of galvanized pipes and the parts below ground level are made from PVC. The absence of dynamic forces bestows an advantage compared to piston pumps. The Rope Pump can be produced with standard materials that are available in any country and with basic tools such as a welding machine and a hand drill. The Rope Pump design is available in the public domain (see www.ropepumps.org and www.smartcentregroup.com).

A number of direct action pumps (i.e. without lever action) generally consist of a moving pipe within a fixed pipe, each having a footvalve. Water flows either up the inner pipe or up the annulus between the two pipes, depending on the design. These include:

- The Canzee pump (Figure 5.3a), which was originally developed privately, although the design was subsequently placed in the public domain. It is used in Angola, Kenya, Malawi, Madagascar, Uganda, Tanzania, and Zimbabwe.
- The EMAS direct action pump (Figure 5.3b), designed and disseminated by Escuela Móvil de Agua y Saneamiento (Mobile School for Water and

(a) The Canzee pump (b) The EMAS pump

Figure 5.3 The Canzee pump and the EMAS pump
Source: (a) Reproduced by permission of Skat Foundation, St Gallen, Switzerland;
(b) Reproduced by permission of Henk Holtslag, SMART Centres, the Netherlands

Sanitation) in Bolivia, and which is used mainly in Central and South America and also in five countries in Africa. Its design is in the public domain (MacCarthy et al., 2013).
- The Tara pump, originally developed in Bangladesh and registered with the Bureau of Indian Standards. It is used in India, Bangladesh, Laos, other parts of East Asia and Africa.
- The Nira AF-85 pump, which is a commercially produced direct action pump manufactured in Finland. It is a more robust version of the Tara. It is used in Ghana, Mozambique, Tanzania, and other African countries.

All four of these direct action pumps combine the use of plastics (often standard pipe sizes and fittings) with steel to avoid corrosion and provide sufficient strength. The absence of a lever and bearings simplifies repairs, when compared to community water supply pumps used at greater depth settings.

Pumps for community water supply (deep well)

If water levels are deeper than about 25–35 m, the human force needed to lift a column of water in the delivery pipe (known as the rising main) soon becomes greater than that which can be exerted without the assistance of a lever, flywheel, or gearbox. Consequently, most of the commonly used community water supply handpumps – outlined below – utilize one or other of these aids (the majority using levers). All use a piston or plunger moving up and down in a cylinder, in combination with a footvalve at the bottom of the cylinder, to draw water in and 'push' it upwards to the surface. It is possible to extend the range of direct action pumps, but only by decreasing the diameter of the cylinder and rising main.

The India Mark II and III, Afridev, and Zimbabwe Bush Pump

The India Mark II and III, the Afridev, and the Zimbabwe Bush Pump are four public domain handpumps that have, according to Baumann and Furey (2013) 'revolutionised rural water supplies'. There is little doubt that the scale of the beneficial impact of these pumps has been huge. Certainly there have been major challenges in their development, manufacture, use, maintenance, and the quality control of replacement parts; and much has been learned in the process. Some institutional aspects relating to these pumps remain controversial to this day; they are discussed further below.

The India Mark II pump (Figure 5.4a) was developed through a major initiative of UNICEF, in partnership with the Government of India and a commercial manufacturer, Richardson and Cruddas, in the second half of the 1970s. This followed, and was part of the response to, a severe drought that hit the Indian states of Bihar and Uttar Pradesh in 1967. One of several objectives of the India Mark II development was to design a robust pump that could be used for a significant period of time before any repairs were needed. Following the early work, important design improvements were made in the 1980s in Coimbatore, India. In the years following initial production, numerous commercial manufacturers produced the pumps in large quantities; the designs were lodged with the Indian Standards Institute (later renamed the Bureau of Indian Standards), which took over responsibility for quality control. The India Mark III, which lent itself much more to so-called village-level operation and maintenance (VLOM), was developed in the mid 1980s. However, it is the Mark II that still dominates the handpump scene worldwide. The maximum lift is 50 m, but at a lift of 25 m the discharge is around 0.25 litres per second. At greater depths, the discharge is correspondingly reduced.

At a similar time, and beginning in Malawi, development commenced of a deep well handpump that could be readily repaired at the village level. This resulted initially in the production of the Maldev pump in 1982. Subsequently, in the early 1980s, efforts shifted to Kenya and, towards the end of the 1980s, 300 Afridev pump units were field tested in Kwale district (now Kwale county). Testing and improvements were also made in Pakistan during this period. The Afridev (Figure 5.4b) differed from the India Mark II in its emphasis on VLOM (because the footvalve, plunger, and pump rods can be removed without dismantling the much heavier rising main) and in its extensive use of modern plastics and rubber. The Afridev is now widely used, second only to the India Mark II in global numbers. At a lift of 25 m

74 RURAL COMMUNITY WATER SUPPLY

(a) The India Mark II

(b) The Afridev

(c) The Zimbabwe Bush Pump

Figure 5.4 Three 'revolutionary' deep well handpumps
Source: (c) Reproduced by permission of Peter Morgan, Zimbabwe

the output is around 0.22 litres per second. The maximum lift by the pump is 45 m, but with a correspondingly reduced output.

The Zimbabwe Bush Pump has a longer history than either the India Mark II or the Afridev. Baumann and Furey (2013) describe its origins in 1933, and its evolution through a number of intermediate stages to something approaching its present form in the 1960s. The Bush Pump was adopted as Zimbabwe's standard handpump, and in the 1980s a complete design review was undertaken by Peter Morgan of the Blair Research Institute. The Bush Pump model B (Figure 5.4c) was adopted as Zimbabwe's standard handpump in 1989. Baumann and Furey (2013) estimate that 50,000 units were in use in Zimbabwe by the time of their publication. Although the Bush Pump can be used at depths up to 80 m (with a correspondingly low output), at a more usual lift of 30–50 m, its output is about 0.4–0.6 litres per second.

The Volanta and the Blue Pump
The Volanta and Blue Pump are both robust, commercially designed and manufactured machines, capable of lifting water from greater depths than the more commonly used India Mark II and Afridev pumps.

The Volanta pump (Figure 5.5a) was developed commercially in the Netherlands by Jansen Venneboer B.V. with support from the Dutch Government around 1980. It is very robust, and only uses corrosion-resistant plastics and stainless steel for components in contact with groundwater.

(a) The Volanta

(b) The Blue Pump

Figure 5.5 The Volanta and the Blue Pump
Source: (b) Reproduced by permission, Skat Foundation, St Gallen, Switzerland

It can lift water from up to 70 m. Two unique features – apart from its large red flywheel – characterize the Volanta pump: first, it incorporates a fine mesh strainer to protect the cylinder, plunger, and footvalve from wear; and second, it uses a stainless steel plunger without wearing seals, thus simplifying maintenance. The Volanta pump has been widely used in West Africa.

The Blue Pump (Figure 5.5b) is a robust lever action pump provided commercially by a partnership between the Fairwater Foundation and the Dutch company Boode. It has a maximum claimed lift of 100 m. The Blue Pump has been installed in Burkina Faso, Cameroon, Central African Republic, The Gambia, Kenya, Namibia, South Africa, South Sudan, and Tanzania.

The Vergnet Hydro pump

The Vergnet pump (Figure 5.6a) is a French-designed, commercially manufactured diaphragm pump. Unlike many of the previously described pumps, the Vergnet transmits hydraulic energy from a piston operated by a foot pedal to an elastic bladder below water. A water-filled pipe connects the near-surface piston to the bladder, and another flexible pipe delivers water to the surface. The pump can be used up to about 60 m lift, although a design variant can extend that lift to about 100 m. The pump can deliver 0.5 litres per second at a lift of 10 m, reducing to 0.3 litres per second at 20 m, and 0.24 litres per second at 45 m.

Mono pumps

Progressive cavity pumps, also known as Mono pumps (after the French inventor Moineaux), utilize a helical rotor, which turns inside a rubber stator that is also formed into a helix (Figure 5.6b). The interaction of the rotor and stator, as the former turns, causes water-filled cavities to progress up the cylinder and subsequently up the rising main to the surface. The pump head has a gearbox allowing a two-handed horizontal axis rotation to transmit a vertical axis rotation to the submerged rotor. Handpumps of this design can lift water from up to 90 m (with correspondingly low discharge), but at the time of writing the non-profit organization Design Outreach is developing the 'LifePump' with the intention of reaching 150 m.

Handpumps – more than technology

All 13 of the pumps described above have their place as part of the physical infrastructure that can supply water to rural communities reliant on fetching water from boreholes or wells. Handpumps appear superficially to consist of simple technology, which can transform the experience of households and communities. However, they introduce additional complications into community life – in particular, how to overcome the challenges associated with their maintenance and repair. Some of these challenges have technical dimensions, but many involve issues of financing and management, which are discussed in subsequent chapters. Here, however, we examine some of the immediate issues raised by the handpump as an option for rural water supply.

78 RURAL COMMUNITY WATER SUPPLY

(a) The Vergnet pump

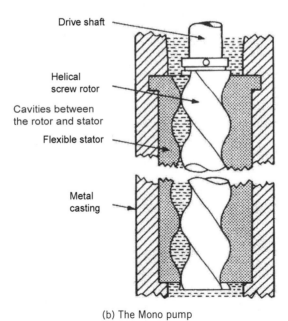

(b) The Mono pump

Figure 5.6 The Vergnet and Mono pumps
Source: (a) Reproduced with permission, Vergnet Hydro; (b) Fraenkel and Thake, 2006

The handpump option – and the issues it raises

In the late 1970s and for much of the following decade (the UN International Drinking Water Supply and Sanitation Decade, 1981–90), a huge initiative was undertaken by a partnership of bilateral and international agencies to identify, catalogue, test, research, and document the state of knowledge of rural water supply handpumps. The main synthesis report of that period was the comprehensive document *Community Water Supply: The Handpump Option*, published in 1987 by UNDP and the World Bank (Arlosoroff et al., 1987). Numerous laboratory and field test reports were also produced, many of which are still available in scanned form. No single body of work on this topic, before or since, has come near to its scope and impact. The combined efforts of many organizations during the water decade of the 1980s led to a number of important concepts and approaches, most of which still form the basis for debate (especially through the RWSN).

Village level operation and (management of) maintenance

The idea of village-level operation and maintenance (VLOM) was first articulated in the early 1980s by the aforementioned UNDP/World Bank Handpump programme. It was primarily a technical 'fix' to an institutional problem – the difficulty of carrying out rapid repairs to handpumps. It was thought that the combination of a sturdy pump with easy-to-make repairs, ideally combined with local manufacture, would lead to shorter downtimes and better service to communities. Indeed, in trials in India there appeared to be some evidence of the success of this attempt to bring repairs closer to the community, reducing the need for and expense of technicians from further afield (Mudgal, 1997). However, as Mudgal admitted,

> The hardware-software divide has been bridged to a substantial degree, and given the right policy environment and institutional capacity building, the management of most of the repairs should be possible at the village level itself. But a more difficult human and institutional challenge persists. The "nuts and bolts" of community participation and management are infinitely more complex to handle. It needs a sustained and determined effort by government, donors, community leaders and most importantly, the users themselves. (Mudgal, 1997)

In the early 1990s, in recognition of shortcomings in VLOM, and also the realization that it is possible for some repair tasks to be delegated by the community to others, the idea of village-level operation and management of maintenance – VLOM(M) – emerged. As Reynolds (1992: 3) put it, 'The VLOM concept was therefore expanded to include "software" or organizational topics. VLOM might now be better understood as standing for "village-level operation and management of maintenance", including: (a) community choice of when to service pumps; (b) community choice of who will service pumps; and (c) direct payment by the community to the maintainers and repairers.' VLOM(M) is now a rarely used term, having been superseded by what for many years has been referred to as community-based maintenance. This is discussed further in Chapter 7.

Public domain vs private commercial

The pattern set in the development of the India Mark II handpump was that the detailed design specification should be in the public domain, with any commercial manufacturer being free to produce and sell it under its official name – so long as the specification was followed to the letter. The same approach was taken with the Afridev, and many other handpump designers have seen the value of placing their specifications in the public domain. The Rural Water Supply Network, following from its predecessor the Handpump Technology Network, has maintained the specifications and documentation of public domain handpumps since 1992, but with diminished resources and little opportunity to be proactive, its effectiveness has reduced.

The public domain design/private commercial manufacture model of handpump production has been controversial, especially among some private companies which have designed and manufactured their own pumps – and who do not wish to share their intellectual property with others. One claim is that once a design is in the public domain it becomes 'fossilized' and further research and development is stifled. In contrast, so the proponents of a more commercial approach argue, the private sector is more likely to continue to innovate and evolve its designs. The main drawback of a fully commercial approach to handpump design and supply, apart from cost, is the vulnerability communities face if or when a particular commercial provider goes out of business, and the supply of pump-specific spare parts dries up.

Standardization

One of the findings of the international work on handpumps in the 1980s was the sheer number and diversity of models in use. Arlosoroff et al. (1987) describe the testing of 70 different pump models, and there were more in use that did not go through the testing programme. The multitude of models was seen as posing a difficulty for countries in regard to assuring supply chains for spare parts and developing technical skills for repair tasks. As a consequence, many countries decided to standardize – that is, to limit the number of government-approved pumps in order to ease their management and maintenance.

MacArthur (2015) provides a comprehensive analysis of handpump standardization in sub-Saharan Africa, finding that out of the 35 countries in the region that use handpumps, formal standardization has taken place in 15, while in all the rest some form of informal (de facto) standardization is in place. She further identifies a number of criticisms of standardization, including the dilemma posed by a government adopting 'the wrong pump' (for example one that is particularly prone to corrosion) or a pump that is generally unaffordable. Nevertheless, with some caveats, MacArthur concludes that standardization in one form or another will remain an important feature of the rural handpump scene for the foreseeable future.

Quality of components

The quality of component parts of handpumps has been recognized from the early work on the India Mark II to be key to the performance of the services provided by the technology. In the 1970s and 1980s formal processes for quality control and assurance were implemented, with support from UNICEF. However, these have gradually weakened, to the point where the quality of spare parts is a major issue for many countries and programmes. The risks inherent in a largely unregulated commercial market have become apparent in the rapid failure of poor-quality replacement parts.

Corrosion

The problem of corrosion of pump components that are in contact with aggressive groundwaters has been known about for many decades (Langenegger, 1994). Casey et al. (2016) carried out a global synthesis and a field investigation in Uganda. In the latter case, and in common with Langenegger's earlier findings, iron concentrations in the 10s of mg/l fell to below 1 mg/l after a sufficient period of pumping and purging of iron-rich water from the pump and borehole. It is apparent that corrosion that takes place over night and at other periods when pumping is not taking place leads to significant build-up of dissolved iron in the borehole water. In many cases the high iron content of drinking water is almost entirely caused by corrosion, as opposed to background levels originating in the natural groundwater chemistry.

Many attempts have been made to design physico-chemical indicators of corrosivity of groundwater, the simplest being pH (in general, waters with a pH<6.5 are judged as corrosive). None of the indicative measures is completely satisfactory, however, and unless extended local experience provides evidence to the contrary, it may be safest to assume that groundwater is potentially corrosive – and to avoid use of iron and galvanized steel components, despite the short-term cost implications. Stainless steel and PVC provide alternatives to galvanized steel.

There are two important consequences of aggressive corrosion. The first is that dissolution of iron from corroding pump components causes unacceptable taste and iron-staining problems for the water users. If these aesthetic water quality issues are particularly bad, users abandon sources, sometimes in favour of alternative sources that may be far less safe from a microbiological (and health) point of view.

The second problem is the damage caused to the pump. In some cases, galvanized steel pipes or pump rods can corrode so severely over a period of just a few months that they become perforated and unable to transmit water (Figure 5.7). Such corrosion makes handpump repair and upkeep unaffordable for the community in question.

Iron or mild steel (even if galvanized) components are especially prone to corrosion, while stainless steel may be less vulnerable; plastics are unaffected from this corrosion. In Uganda, the Ministry of Water and Environment has responded to the evidence of corrosion studies, such as those cited above, by no longer permitting the use of galvanized steel rising mains and pump rods.

Functionality

Handpump functionality has become a topic of wide interest, especially since the RWSN published its estimates for 20 countries in sub-Saharan Africa in 2009. In their short note (RWSN, 2009), republished in a more substantive paper a year later (RWSN, 2010), the country statistics of handpump functionality were the estimates or best guesses of informants from whom 'data' were collected. The statistics varied widely in their rigour and quality, but the conclusion that 'many [around 10 to > 65 per cent] handpumps, considered a robust and simple to maintain option, are actually out of operation' has been very influential. The widely quoted average of 60–65 per cent functionality has been taken to mean that 35–40 per cent of handpumps are permanently out of service. As I argue below, this is almost certainly not true.

More recently, the statistics on water point (not only handpump) functionality have been updated using the data in the water point data exchange (WPDx, n.d.). Banks and Furey (2016) estimated that an average of 78 per cent of water points were functional across the 11 countries for which there were sufficient data for analysis. However, they pointed out that

> the high failure rates early after installation are troubling: almost 15% after one year and 25% of water points are non-functional by their fourth year after installation. This indicates widespread problems with poor quality water point installation, due to a range of problems that may include professionalism and skills around contracts, construction and supervision; borehole siting; lack of quality control of hardware; or lack of post-construction monitoring and problem resolution. (Banks and Furey 2016)

Some of these issues have been highlighted already in this book, and are discussed further in subsequent chapters.

A final conclusion by Banks and Furey is that functionality statistics for handpumps are little different to those of other technologies. A knee-jerk reaction to the apparently poor functionality of handpumps might result in a preference for alternatives that may serve their users no better. Perhaps it is the case that no particular technology is inherently better or worse, but rather the way the water supply is implemented, managed, and financed is more important. These matters, too, are the focus of subsequent chapters.

WATER LIFTING FROM WELLS AND BOREHOLES: HANDPUMPS

Figure 5.7 Corrosion of India Mark II galvanized rising main in northern Uganda

A hidden crisis

A major research project conducted under the UPGro (Unlocking the potential of Groundwater for the Poor) programme investigated the functionality of boreholes equipped with handpumps (HPBs) in Ethiopia, Malawi, and Uganda. The 'Hidden Crisis' project was active between 2013 and 2020, and it published numerous reports and papers.

The project brought together teams focusing on the natural science and technology of HPBs and their functionality, and other researchers who focused on the social and management dimensions of the functionality problem. New approaches to the definition and measurement of functionality, and the assessment of the adequacy of water management arrangements, were developed (Bonsor et al., 2018; Fallas et al., 2018).

The main conclusions of the project can be summarized as follows. Using the tiered definition of functionality developed by the project, the country statistics were as set out in Table 5.2. In the table, the successive columns are cumulative, for example the figure of 28 per cent for Ethiopia (top row, right) refers to the percentage of water points that were working, *and* delivering at least 10 litres per minute, *and* with a down time of no more than 30 days in the last year, *and* supplying water compliant with WHO quality standards.

1. In Ethiopia and Malawi, the water management arrangements were found to be 'functional' or better at 85% and 86% of sites, respectively; in Uganda (the country with the lowest physical functionality scores), these arrangements were found to be weak at 70% of sites.
2. More detailed surveys of a sub-set of the HPBs examined hydrogeological and groundwater engineering aspects. The findings are summarized in Kebede et al., (2019), Mwathunga et al., (2019), and Owor et al., (2019).
3. The project found only weak relationships between the capacity of the water management arrangements and the resulting functionality scores (Whaley et al., 2019).
4. It also highlighted the limited ability of local governments in all three countries to provide support to community-based maintenance arrangements (Le Sève, 2018; Oates and Mwathunga, 2018; Pichon, 2019), despite this aspect being found to be relatively strong in the community-based social surveys (Whaley et al., 2019).

Table 5.2 Country statistics for functionality

	Working	At least 10 litre/min	Down time no more than 30 days in last year	WHO standard water quality
Ethiopia (n = 172)	82%	59%	45%	28%
Malawi (n = 200)	74%	67%	58%	41%
Uganda (n = 200)	55%	34%	23%	18%

Source: Kebede et al., 2017; Mwathunga et al., 2017; Owor et al., 2017

5. The project explored the impacts of down time and poor functionality on water users, finding that communities routinely experience high levels of water stress. These include both regular pressures from funerals, cultural events, and dry periods; and cascading pressures as a consequence of routine sharing of water points with neighbouring communities due to poor functionality.

Beyond functionality

The notion of functionality has been criticized. Carter and Ross (2016) set out seven reasons why functionality as an indicator is less than ideal. Three of those seven reasons are especially pertinent. First, however it is defined and determined, functionality is a 'snapshot' measure; a water point that is working today may not be working tomorrow (and vice versa). Consequently, functionality tells us little about communities' struggles with, and successes and failures over, water point management.

Second, functionality is usually reported for a geographical or administrative area, or for a group of water points that are the responsibility of a single local government or non-governmental organization. It is reported as a ratio or percentage (the number working divided by the total number). The difficulty often lies with incorrect or incomplete data regarding the second figure (the denominator). If a significant number of water points have been decommissioned, abandoned, and/or forgotten, then this figure will be underestimated, and functionality will be overestimated.

Third, and most important, functionality is not the same as sustainability. Carter and Ross (2016) set out the skeleton of a research agenda:

> If we are trying to understand how non-functionality actually impacts on services and people, then standard measures such as mean time between failures … and down times … are far more useful. Furthermore, knowledge of the proportion of water points of any given age which are abandoned, and the relation of this parameter to age, is crucial. Were we to have data on these three parameters, we would have a much better quantitative understanding of functionality and ultimately of service performance. To complement such quantitative data, we need better narratives, histories of water-point breakdown, of struggles to raise funds, and of successes and failures in regard to water-point repair and maintenance. (Carter and Ross, 2016: 108)

These issues are explored further in Chapter 7, where a particular emphasis is placed on identifying preconditions for successful water management rather than causes of failure of failed or failing services.

The place of handpumps in the last decade of the SDGs

Handpumps and the services they provide to water users will continue to be needed for at least the following few decades, unless and until piped water reaches everyone – a distant if not impossible prospect for many remote and low-density populations. Some countries have already transitioned out of handpumps; others are in the process of doing so; but for a significant number, where many of the poorly served rural populations live, the humble handpump will be needed for a while longer (Carter, 2015).

Given the sheer number of handpumps 'in the ground' (several million) and the number still being installed each year – around 100,000, including 60,000 in sub-Saharan Africa (Sansom and Koestler, 2009) – it is apparent that rural water supply handpumps need an adequately resourced and independent 'champion', as MacArthur (2015) argues. Such an organization, convened at a global level, would not only provide guidance about standardization, but also support quality control and assurance, evaluate design modifications, and update specifications and standards. Unlike other technologies that are perceived as more modern and attractive, there is a real gap here; were such a champion to emerge (or the existing one be adequately resourced), this could have a major beneficial impact on many of those who form the focus of this book – the rural poor.

Energy and power for water pumping in rural areas

The basic household water needs (of about 20 litres per person per day) of a household of five persons can be met by pumping 100 litres of groundwater. The energy required can be provided by conversion of the energy in food (in the case of manual pumping), from fossil fuel (in the case of a petrol or diesel driven pump), or using electricity provided from the grid (if available) or from a solar photovoltaic array.

Clearly there are significant benefits to be gained if water can be lifted mechanically to an overhead reservoir, and thence be distributed by gravity to individual homes or to public standposts. Given the major drawbacks of pumping using fossil fuels, and given the common absence or low reliability of grid electricity in rural areas, the first part of Chapter 6 is devoted to solar water pumping.

CHAPTER 6
Water supply infrastructure: beyond handpumps

Abstract: *This chapter examines some of the technical, managerial, and financial implications of moving 'up the ladder' of rural water supply. It outlines features of piped and mechanically pumped systems, which can potentially enable higher levels of service, up to and including individual house connections. Gravity flow schemes are particularly attractive, where water resources, topography, and settlement patterns permit. However, even the infallibility of gravity is insufficient if management systems and recurrent finance are inadequate. Mechanical pumping of water, using direct drive from diesel engines, grid electricity, diesel generators, and solar pumping all have advantages and disadvantages, but solar pumping is of particular interest nowadays because of its much-reduced capital costs. Nevertheless, recurrent costs and management challenges are ever-present. In addition, all piped systems require adequate water storage capacity, and this represents a significant cost component. Design of piped systems follows well-established procedures, and water may be delivered to public standposts, kiosks (for onward sale), or to individual properties. It is important to set appropriate charges and tariffs that reflect the different levels of service enjoyed by those with house connections as compared to users of public taps.*

Keywords: gravity flow schemes, mechanical pumping, solar water pumping, piped distribution systems, public standposts, kiosks, house connections

> 'Just remember, Callum, when you're floating up and up in your bubble, that bubbles have a habit of bursting. The higher you climb, the further you have to fall'
> —Malorie Blackman, 2006

Introduction

Wells or boreholes equipped with handpumps are representative of the minimum physical infrastructure needed to provide basic water supply services. Such minimum infrastructure also includes protected springs (without piped distribution of water) and rooftop rainwater harvesting systems, if they can provide year-round drinking water. For many communities, such a level of technology is what they aspire to, since even this basic infrastructure has not yet reached them. For others who do enjoy this level of infrastructure, there may be little prospect of further progress in the short- to medium-term – for reasons of affordability (by household or government) or the constraints imposed by limited management capacity.

For another segment of the rural population, a more sophisticated range of pumped and piped water services is possible, enabling access for at least some people to piped house connections. Many national governments and international organizations believe that such an achievement is possible for all people by 2030, as expressed by Sustainable Development Goal 6.1. I do not share the optimism over the time scale, although I concur with the ambition to work towards the provision of reliable, affordable, and safe piped water over whatever time period is realistic for each country.

It is against this background that this chapter considers a range of gravity-fed and pumped, piped water supply systems. Such systems may be the best or only option in some circumstances; in others, they represent alternatives to point sources, and decisions have to be taken about which alternative is the preferred option.

A word of caution at the outset of this chapter: mechanically pumped and piped water systems, and even gravity-fed piped systems, are generally more expensive to construct and run, and they are more complex to manage, than point sources. Higher levels of service come at a financial and management cost. This is discussed further in Chapters 7 and 8.

Gravity flow schemes

If a large enough spring or uncontaminated stream is available at a higher elevation than the users it is to serve, then water may be delivered via a pipe network using gravity alone as the driving force (Figure 6.1). The simplicity of such gravity flow schemes (gfs) is beguiling. While pumps and their energy sources are prone to breakdowns, gravity is utterly reliable. Unfortunately, as we shall see, gravity flow water supply schemes are not without their reliability challenges.

The main components of a simple gfs are as follows:

- the source (usually a protected spring or an offtake from a stream);
- a service reservoir, which can receive a steady inflow while delivering a variable outflow during the day;
- a piped distribution system;
- public standposts, kiosks, and individual house connections.

Other features such as sedimentation tanks, break-pressure tanks, air valves, and additional local service reservoirs may be needed in specific cases.

The design of gravity flow water supply systems requires careful consideration of future water demands, knowledge of the topography of the area, and understanding of pipe hydraulics. Water quality must be considered, especially if there are risks of corrosion, biofilm development, or encrustation. Good practical design guides are readily available (Jordan, 1984; Arnalich, 2009, 2010), and open-source software is readily available to aid in the design process, the most widely used being EPA-Net (Arnalich, 2011a, 2011b).

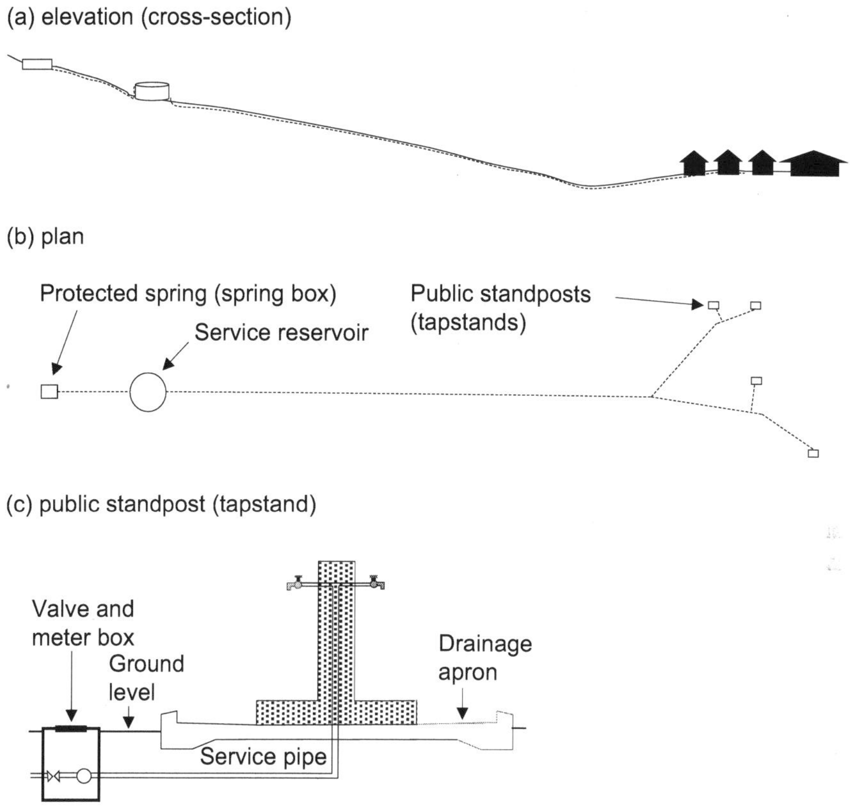

WATER SUPPLY INFRASTRUCTURE: BEYOND HANDPUMPS 89

(a) elevation (cross-section)

(b) plan

Protected spring (spring box)
Service reservoir
Public standposts (tapstands)

(c) public standpost (tapstand)

Valve and meter box
Ground level
Drainage apron
Service pipe

Figure 6.1 Simple branched gravity flow schemes
Note: Dashed lines signify buried pipelines

The difficulties faced even by gravity flow schemes are well illustrated by the experiences of the Malawi Rural Water Supply Program, which ran from 1980 until 1986, building on substantial earlier work. The programme was responsible for constructing (or in one case reconstructing) 19 rural piped (gravity) water schemes. The projects were constructed to serve a design (future) population of 422,000, although Kleemeier (2000) estimated that nearer three times this number would have been served by the longer period of investment in gravity flow schemes.

At its end of project evaluation the project was described as 'probably the most outstanding rural piped water program in Africa'. The success of the project was attributed to 'the full involvement of the user communities, field staff within the Ministry of Works and Supplies (MOWS) and Ministry of Health (MOH) ... and dedicated senior staff within the MOWS and MOH' (Warner et al., 1986).

Just over a decade later Kleemeier (2000, 2001) undertook various studies of the project. She cites no fewer than 20 documents that, like the final evaluation, praised the project in the highest terms. And yet she found that 'between three and 26 years after completion, the smallest schemes, and the newest one, are performing well, but about half the schemes are performing poorly, and a third of these are functioning abysmally' (Kleemeier, 2000: 929). Kleemeier placed the blame for this mostly at the door of the local government system, which was supposed to support community management. Local government in Malawi has been consistently under-resourced and consequently unable to provide the support needed to effectively manage larger gravity flow schemes. Kleemeier concludes that 'when local organizations have links to an ineffective administration, they can manage only the simplest types of technology' (2000: 942). Included in those simple technologies are the smallest gravity flow schemes, but not schemes covering many kilometres or tens of kilometres. In short, even the reliability of gravity itself, and the apparent simplicity of piped delivery, are no panacea – effective, coordinated management and adequate post-construction financing are also vital.

Mechanically pumped systems

Going beyond point sources to piped supply poses additional financial and management challenges to those that are already inherent in point sources, such as boreholes equipped with handpumps. Going beyond handpumps to mechanical pumping represents a further difficult step.

Mechanical pumps can use grid electricity if it is available and reliable. However, many of those who lack adequate domestic water supply services also lack grid electricity. This is especially true in rural areas of low-income countries. In the absence of reliable grid electricity, pumps may be driven directly using diesel engines (Figure 6.2a) or by using diesel-powered generators (Figure 6.2b). Both options have their pros and cons, but neither is ideal because of the requirement for steady supplies of diesel fuel, the price volatility of that energy source, and of course the adverse environmental impacts of burning fossil fuels. Repairs and maintenance also require reliable service from competent specialist technicians.

Electric submersible pumps

Before examining the solar water pumping option, it is important to consider first the technology of electric submersible pumps, as these are used with grid mains supply, generator-produced electrical power, as well as solar systems themselves.

A typical electric submersible pump (Figure 6.3) consists of the following components:

- the electric motor, supplied with power via a cable which is strapped to the rising main;
- the strainer, where water enters the pump;
- the pump itself, usually consisting of multiple stages or impellers.

(a) Diesel engine (left) supplying power to a borehole pump via a belt drive (right, in cage), Tanzania

(b) Diesel generators supplying electrical power for high-lift pumping, Ethiopia (left) and Timor-Leste (right)

Figure 6.2 Mechanical pumping infrastructure

The submersible pump hangs in the borehole casing, below water. It should be a fairly close fit within the casing (and not within the well screen). This is to ensure that water entering the borehole through the screen (which is usually below the pump) flows at high velocity up the annular space between motor and casing (before entering the strainer), so keeping the motor cool. If a submersible pump is used in a large diameter well, then it should be covered in a metal shroud to achieve the same effect.

When electric motors start up, they draw significantly more current than when they are running under normal load. Consequently, decisions regarding wiring, generators, and other electrical components need to take this into account in the design process. Submersible electric pumps must not be allowed to pump dry. To avoid this, it is common to incorporate a water level sensor above the pump (or some other arrangement to achieve the same end), which automatically cuts the power supply if the water level falls too close to the pump.

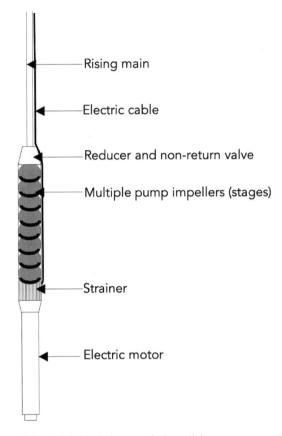

Figure 6.3 Submersible electric borehole pump (schematic)

Each submersible pump has its pump characteristic curve (the relationship between head and discharge), as well as its relationships between discharge, efficiency, and power requirements. The nature of these relationships is indicated in Figure 6.4. The head and discharge required from the pump are determined by the design of the water delivery system. The relationship between head and discharge for the piped system is known as the system curve. As discharge increases, pipe friction head losses also increase, so the system curve rises as shown by the dashed line crossing the pump curve in Figure 6.4. The point at which the system curve crosses the pump curve is the duty point at which the pump will run. It is important that this duty point enables the pump to operate close to its peak efficiency, since anything less implies a continuous unnecessary energy cost.

Pump selection and specification requires consideration not only of the points just made, but also the amount of submergence needed (the required net positive suction head), the power – especially the starting power needed – and the water level cut out arrangements that are desirable.

WATER SUPPLY INFRASTRUCTURE: BEYOND HANDPUMPS   93

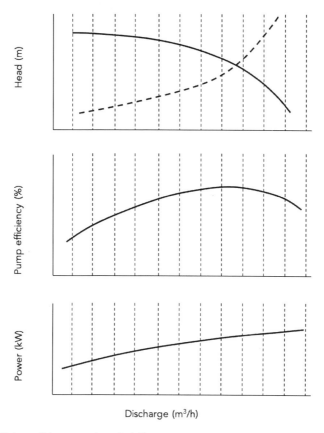

Figure 6.4 Submersible pump characteristic curves
Note: The dashed line in the top graph is the system curve.

Solar water pumping

It has only recently become relevant to discuss solar water pumping in the context of rural water supply because of the sharp fall in the cost of solar photovoltaic (solar pv) technology in the 2010s. There is a two-way link between the price of the technology and the size of the market (or the installed capacity). Over the period since the late 1970s (when prices were prohibitively high), the price per Watt peak (Wp, see below) has fallen by a factor of more than 100. The price (at the time of writing) of electricity generation by this means is less than US$0.75 per Wp (World Bank, 2018), and solar pv can increasingly compete with other means of generating electricity.

Standard solar water pumping systems

A standard solar water pumping system consists of an array of modules or panels, which are exposed to direct or diffuse sunshine, thus producing electricity. A control box placed between the solar array and the pump motor ensures that

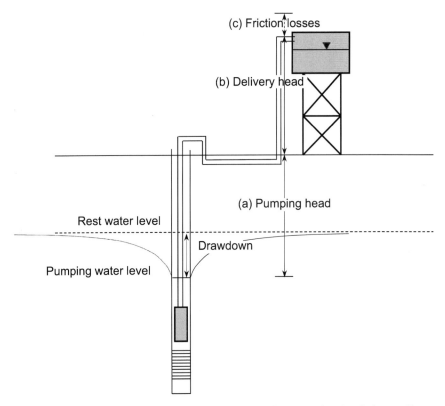

Figure 6.5 The total dynamic head (a+b+c) for a submersible pump in a borehole or well

the electricity supply to the motor is correct. The most common pump type used is a submersible electric unit, as just described. The pump delivers water to a raised tank (Figure 6.5) from which it is distributed by gravity to any or all of individual household connections, public taps, or water kiosks.

Solar pv electricity generation systems are rated according to their output under peak solar radiation conditions of 1000 W/m^2, and at a standard temperature of 25°C. Solar radiation intensity varies continuously through the day, and array temperatures at times of peak radiation may considerably exceed 25°C. Consequently, it should be borne in mind that a solar module or panel rated at, say, 100 Wp may only produce 100 Watts of electrical power for a short part of each day.

Solar pv arrays produce direct current (DC). If the pump has a DC motor, then the extent of power conditioning by the control box is limited, with correspondingly small energy losses. If alternating current (AC) pumps are used, there is then a requirement for an inverter to transform DC to AC. Inverters add further cost and energy inefficiency to the system.

The pump must be capable of delivering the required discharge at the total dynamic head (Figure 6.5), which consists of three summed components:

> **Box 6.1 Head and the hydraulic grade line in piped systems**
>
> The concept of head is fundamental in all systems that contain water, whether static or flowing. These include surface water and groundwater, and piped systems. The total head of water (its inherent energy) is made up of three components, each of which can be expressed in metres of water:
>
> - the elevation head (i.e. the height of the location above a chosen datum – for example, sea level);
> - the pressure head (equivalent to the height of water above the point in question);
> - the velocity head, a measure of kinetic energy, calculated as $v^2/2g$, which is generally negligible at the velocities (v) with which water flows in piped systems.
>
> In simple terms, therefore, head in a piped system is the sum of elevation and pressure heads. There is then one unbreakable rule – that water always flows from higher to lower total head. It does not necessarily flow from a higher point to a lower point (otherwise it could not be delivered to the top of a multi-storey building); it does not always flow from higher to lower pressure head (otherwise water at the bottom of a tank – where pressure head is higher – would constantly be flowing upwards to the top of the tank).
>
> The total head at a point in a pipe system is the height (above the chosen datum) to which water would rise in a vertical pipe attached to the pipe at that point. In designing the pipe system, the hydraulic grade line (the imaginary line joining those values of total head) must always be kept above the minimum residual pressure head. Pipe sizes are chosen accordingly. Figure 6.7 shows the hydraulic grade line under no-flow and flowing conditions in a simple pipe system.

(a) the pumping head (i.e. the vertical distance from the pumping water level to ground level, during drawdown in the well or borehole); (b) the delivery head (i.e. the additional vertical distance to the delivery point at the top of the raised tank); and (c) friction losses in the rising main and any fittings (which are a function of the flow rate). See Box 6.1 for a further discussion of 'head'.

The storage reservoir normally needs to have a capacity equivalent to 2–3 days' water supply, unless there is a backup pumping option for those periods when cloud cover is heavy or when repairs need to be carried out.

The size of the solar array and the specification of the pump and motor are determined by the required total dynamic head and discharge (both of which may vary through the year), together with the available solar irradiance at the location in question (which is everywhere variable across the seasons). Standard calculation procedures are used in system design, as outlined by the World Bank (2018); the major equipment manufacturers and suppliers have their own design procedures and software.

Wider issues concerning solar pumping

Cost. Although the single most significant factor in driving prices down has been the increased efficiency and reduced cost of solar modules, the array only represents one element of a rural water system. The full list of major cost

Table 6.1 Water delivery system components compared

Solar-pumped	Handpumped
Borehole	Borehole
Electric submersible pump	Handpump
Solar array	N/A
Power conditioning	N/A
Cabling	N/A
Water storage	N/A
Reticulation system	N/A

components is as set out in Table 6.1, side-by-side with a comparison for a handpump installation.

The World Bank (2018) gives an example of a rural community solar water-pumping scheme serving 2,000 people (at 30 litres per person per day), and supplying water over a total dynamic head of 130 m. Such a scheme would require an array size of just over 10 kilowatts peak (kWp) and a water storage tank of 120–80 m³. The authors estimate that the total installed price would be around $32,900 (i.e. about $3.2 per Wp or $16.5 per person). They point out that this price would apply to larger systems (>10kWp) or multiple smaller systems, where economies of scale can be achieved. It is important to note, however, that the costs of the borehole, the storage tank, and the reticulation system are not included in this analysis; the actual capital cost is therefore likely to be around double the figure given – about $33 per person for the physical infrastructure alone.

A more wide-ranging cost analysis was undertaken by Armstrong et al. (2017), who examined records from 85 schemes implemented by Water Mission (a US-based non-profit organization) of various sizes in eight countries. They determined the ranges of capital expenditure (capex), operation and minor maintenance expenditure (opex), and capital maintenance expenditure (capmanex), following the analytical framework of Fonseca et al. (2010; see also Chapter 8). Table 6.2 summarizes their findings.

A few points are worth noting in this analysis: first, the average design water quantity per person was less than 10 litres. These schemes were

Table 6.2 Capex, opex, and capmanex costs of solar pumping schemes of different sizes

Cost component (average US$ per person)/ Scheme size	Small <500 people	Medium 500–5,000 people	Intermediate 5,000–15,000 people
Capex	140	39	10
Opex (per year)	1	0.5	0.1
Capmanex (per year)	–	1.5	–

Source: Armstrong et al., 2017

designed to provide chlorinated water, primarily for drinking and other consumption purposes. Second, nearly 90 per cent of the schemes analysed were in the small or medium category, serving up to 5,000 people. Third, the average capex cost (for medium-sized schemes) works out nearly double that summarized in the World Bank example above, probably reflecting Water Mission's inclusion of all physical and software costs (the latter adding about 33 per cent to the physical costs). The typical array size in Armstrong et al.'s analysis was much smaller than the World Bank example, at 1.5–2.0 kWp. Finally, in the estimation of capmanex, the assumptions made about estimated lifespans of various components (none less than 10 years, some up to 30 years), the inflation rate used (2.8 per cent p.a.) appear, to this author at least, somewhat optimistic. It is only if components, materials, and construction quality are of the highest order and preventive maintenance is rigorously practised that solar water systems will prove their sustainability.

In round figures, the capex and opex costs of a handpump service to a small community are up to $50 per person and $0.50 or less per person per year, respectively. This means that solar pumping only becomes financially attractive for larger communities of at least 500 people, and typically 2,000 or more. At this scale, the yield of groundwater sources may become an issue, thus requiring more boreholes, so increasing the investment (capex) cost. Furthermore, the capital maintenance costs more generally may well turn out significantly higher than Armstrong et al.'s analysis suggests, as the quality of components and installations undertaken in the absence of the close supervision that Water Mission can provide may be poorer.

To summarize: the actual costs of solar water pumping systems are very dependent on (a) the quality of the installations; and (b) economies of scale, achievable by serving relatively large communities and installing many schemes in a geographical area. A key factor in financial sustainability is the actual capmanex costs. Major breakdowns due to poor quality equipment or installation, damage to the system, or theft of equipment would necessitate costly replacements, so pushing capmanex costs higher than those indicated here.

Complexity and management. Despite the World Bank's enthusiasm about the potential of solar water pumping, they acknowledge that 'sustainability of solar water pumping has been a challenge in many countries and especially in rural areas', which they attribute as 'due to lack of proper O&M' (World Bank, 2018: 29).

There is little doubt that the maintenance and repair of solar water pumping technology generally requires a more advanced set of skills than those which can readily be transferred to community members in lower-income countries. Consequently, community management of solar water pumping systems may be unrealistic. The World Bank recommends that communities establish 'comprehensive maintenance contracts with suppliers during warranty periods and ... beyond' (2018: 29). Such contracts of course come at a cost, which, under the present 'consumer pays' philosophy, would have to be borne in full by water users. Later in this book I question the principle that the rural consumer should pay the full lifecycle cost of a water supply service; this principle is not applied in urban settings, nor in the case of other public services.

A new entrant: the Impact Pump

Concerns about the true post-installation costs and the necessary management requirements of water lifting by conventional solar pumping have led a UK-based company, Thermofluidics Ltd, to develop novel technology in the form of the Impact Pump (https://www.impactpumps.com). This pump uses the same principle that is used by hydraulic ram pumps, but in the form of a down-hole unit driven by water pumped from the surface. Hydraulic ram pumps in conventional surface applications can be highly reliable (Fraenkel and Thake, 2006).

The Impact Pump, installed in a borehole or dug well, is driven by water pumped down a flexible pipe using a small solar-powered pump at the ground surface. The supply of water in this way, combined with the shock waves generated by the pump, allows it to raise up to the same volume of water again, over up to twice the head-difference across the surface-level pump (Figure 6.6).

The considerable research and development efforts that have led to the Impact Pump have focused on: (a) keeping the (capital) cost low; (b) achieving high efficiency at water lifts of up to 25 m and outputs of up to 0.55 litres per second (for the initial product); and (c) ensuring maximum reliability. The manufacturer provides a full 5-year warranty on the pump, without standard exclusions (e.g. related to water quality). This is considerably longer and more broad-ranging than the 1–2 year defects periods provided by conventional submersible pump manufacturers and/or suppliers (World Bank, 2018).

Because of the high reliability of the Impact Pump, maintenance and repair tasks focus on the components of the system that are above ground and easily accessible, namely the drive pump (which can be any suitably sized unit, with any chosen energy source), the storage tank, and any delivery pipework.

At the time of writing, the Impact Pump is in manufacture, with imminent rollout across a number of African countries. Its performance and sustainability will be a good test of the potential for technical innovation to overcome, even in part, the management and financing challenges of sustainable rural water supply.

Solar water kiosks

In the 21st century a number of commercial companies and non-profit organizations have designed standalone solar water kiosks to supply

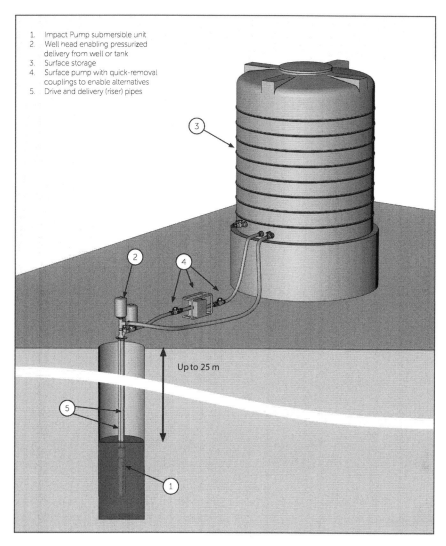

Figure 6.6 The Impact Pump

potable drinking water. These installations generally consist of some or all of the following components:

- a borehole or surface water source with a submersible pump;
- a solar array with cabling and electronic control systems;
- some form of advanced water treatment (such as reverse osmosis);
- an elevated water storage tank supplying water to user collection points;
- sensor technology to detect failures and enable rapid repairs;
- payment technology, utilizing prepaid tokens or mobile phone payments;
- a business or social enterprise arrangement addressing responsibilities for operations, preventive maintenance, repairs, and financial management.

There is little doubt that such systems can perform effectively (in terms of delivering safe and reliable drinking water), but as with all physical infrastructure, the performance of the management and the adequacy of the financial arrangements are crucial. The water produced by such systems tends to be expensive, so raising questions about affordability, inclusion, and financial viability.

As an evaluation by Pan-African consulting firm EED Advisory (2018) has pointed out, theft, vandalism, and component failure are to be expected; how the community and its supporting institutions (or the private operator) respond, and whether they have the funds to undertake rapid repairs, are the critical issues (see Chapters 7 and 8).

Water storage

All piped systems require service reservoirs (water storage tanks), the purposes of which are to:

- enable a steady or varying inflow of water (either continuously over 24 hours, as from a spring or surface water diversion; or a steady flow over several hours of the day, as with a fossil fuel powered pump; or a variable flow in the case of solar pumping) to match with the highly variable demands for water from consumers;
- provide elevated storage from which the community can be supplied by gravity;
- provide a volume of water that is available for use if the pump breaks down or is out of service for planned maintenance; or, in the case of solar pumping, to allow for periods of prolonged low irradiance due to cloud cover;
- provide storage for emergency use, for example in firefighting.

Water demand by small community water consumers is highly variable over the course of a 24-hour period, with the demand in the peak hour (often the early morning) commonly three to four times the average rate over the day. Night-time demand falls to very low levels.

If the expected demand pattern can be realistically assumed on an hourly basis – and if the (steady or variable) inflow rates to a planned reservoir can be predicted, also hourly – then it is easy to calculate the volume of storage needed to fulfil the first of the purposes listed above. The technique involves either tabulating or plotting these two hourly data sets, and examining the cumulative differences between them (Wagner and Lanoix, 1959). The volume of storage calculated in this way is typically one-third to one-half of a day's supply. Such a storage capacity represents the minimum needed, with no allowance for continuation of supply when, for whatever reason, the inflow reduces significantly or ceases completely. Nor would it allow for emergency uses such as firefighting.

In the field of solar-powered drinking water supply, a range of different practices is followed. The Practica Foundation, for example, recommend sizing the water storage tank on the basis of 40–60 per cent of a day's supply to avoid what they consider to be otherwise excessive costs (van de Giessen, n.d.). They recommend increasing the array size rather than increasing water storage capacity to compensate for days with low solar irradiance – although this does not allow for pump or critical component breakdowns. The World Bank (2018), on the other hand, suggests two to three days' storage capacity as the norm. The cost differences between these alternatives are very significant.

Service reservoirs may be constructed of plastic (usually high-density polyethylene, HDPE), unreinforced cement mortar, ferrocement, reinforced concrete, or steel; if they are elevated they may be mounted on steel or reinforced concrete towers. Reservoirs, especially elevated tanks, are expensive. Cost per cubic metre of water stored reduces with size in all cases (Parker et al., 2013), but their significant absolute cost, regardless of capacity, means that they form some of the most expensive major components of water supply systems in which they are needed.

Piped distribution systems

Downstream of the main service reservoir, the pipe network is designed to ensure that there is sufficient pressure at all necessary points in the system. This in turn means that the short service pipes (Figure 6.1) supplying water to public standposts, kiosks, and individual properties can deliver the required flows. The design process consequently proceeds as follows:

1. An initial sketch of a suitable pipe network (either a simple branched system or a more complex looped or gridded network) is laid out on the topographic map of the area; this also identifies the desired locations of public standposts or kiosks and other locations from which service pipes are to deliver water to, for example, schools and health care facilities. The layout should attempt to avoid high points (where airlocks can form) and low points (where sediment can accumulate). Unnecessary crossings of watercourses, and other practically or logistically difficult locations, should also be minimized.
2. At each delivery point, a design discharge is determined; this should be the anticipated peak hour discharge. The typical design discharge for an individual tap would be about 0.2 litres per second.
3. A minimum residual pressure is allowed for at every point where a service pipe will branch off the main system. The residual pressure has to overcome (a) any elevation difference between the junction and the outlet, (b) the friction loss in the service pipe, plus (c) the head needed to deliver the water through the outlet itself (negligible in the case of an open tap, but significant in the case of a ball valve controlling the flow into a water tank). If water is to be delivered to a public tap, a few metres from the junction and at a similar or lower elevation, then a residual

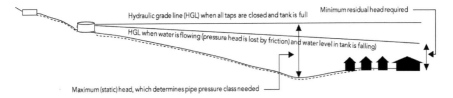

Figure 6.7 Head and hydraulic grade line in a simple piped system

pressure of 5 m is normally sufficient. If, however, water is to be supplied over a long distance or to a higher elevation (such as a multi-storey building or an overhead tank), a higher pressure is needed. Arnalich (2010) recommends a minimum residual head of 10 m in all cases.

4. Once the first sketch of the network has been established, identifying the elevations of all the nodes (junctions with service pipes) and the required outflows, the required pipe diameters can be selected.

A fundamental principle in pipe system design is the concept of head, and the related idea of a hydraulic grade line. These concepts are described in Box 6.1 and Figure 6.7.

In selecting the required pipe diameters, the key relationship is that between pipe material, pipe diameter, and head loss as a result of friction. Head losses due to friction are generally roughly proportional to the square of the flow velocity or discharge; they are directly proportional to the length of pipe; and inversely proportional to the fifth power of diameter. In other words, a doubling of discharge would result in the head loss increasing by a factor of nearly four; a doubling of length would result in a doubling of head loss; and a doubling of diameter would mean that head loss is reduced by a factor of about one divided by 32, or about 0.03.

Head losses due to friction can be calculated either using established semi-empirical equations, such as the Hazen-Williams formula, using published pipe friction charts (Arnalich, 2009), or using open-access or commercially available software (Arnalich, 2011a, 2011b).

The outline presented above covers the basics of piped system design. Many other factors need to be considered however, including:

- the selection of pipe materials and necessary pressure class(es);
- requirements for losing excess pressure (achieved by the use of break-pressure tanks);
- arrangements for road, river, or gully crossings, and pipe anchoring (at bends);
- requirements for air relief valves at high points and washouts at low points and dead ends;
- pipe burial, marking, and documentation.

Further details of these matters are addressed in Arnalich's very practical publications (2009, 2010) and other standard water supply texts.

(a) Public standpost, Uganda   (b) Water kiosk, Malawi

Figure 6.8 Public standposts and water kiosks

Public standposts, kiosks, and house connections

The final destinations of water in piped systems are the outlets, in the form of public standposts or tapstands (Figure 6.1c and Figure 6.8a), kiosks (Figure 6.8b), and individual house connections. Even such apparently simple physical structures require periodic maintenance and repairs. Concrete work needs repair, and taps (especially cheap but readily available taps) require regular replacement.

It is common in systems having a mixture of public taps and house connections that households be required to pay to be connected directly to the system (to cover the materials and labour costs of the service pipe and water meter), and a different tariff than that charged to public tap users. Often those using public taps pay by volume at the time of collection (a charge per jerry can or bucket), while those with house connections may pay on receipt of a water bill. Despite the greater usage by those with house connections, the volumetric tariff they pay is not always higher, and in some cases it is lower – meaning that those with an inferior level of service may be subsidizing those enjoying piped water in the home.

CHAPTER 7
From getting it going to keeping it flowing: management of rural water services

Abstract: *A focus on access to water should not distract attention from the equally important matter of keeping services working. Regardless of who manages rural water services, a common set of key attributes and capabilities is needed. Self-supply is a natural approach to both provision and subsequent management, since water points are constructed and managed by individual households. Externally implemented community water supply programmes will continue to be needed, and their scale and pace of development must increase, but subsequent management by communities is not easy. Much greater attention needs to be paid to improving the effectiveness of community management. Other models for rural water supply management do exist. While management by non-specialized local governments and municipalities is generally ineffective and should not be considered, management by specialized public utilities, and by mandated private operators, can be effective in situations where implementation is possible. Community management will remain the dominant model for some time to come; evidence from low- and middle-income countries shows that it can be made to work, provided that well-established principles are adhered to, including effective budgeting for support.*

Keywords: rural water management, management arrangements, self-supply, community management, utility management, private operator management

'We've only just begun'
—The Carpenters, 1970

Introduction

All initiatives that aim to enhance aspects of rural water services involve an initial investment of resources and effort to bring about beneficial change – getting it going – followed by establishing the necessary arrangements for the management and financing of the service for the foreseeable future – keeping it flowing. This chapter addresses both parts of that timeline, but focuses especially on the second. I begin with the first step, however, because how a new water service is established in the first place has implications for how it can be effectively managed thereafter.

Since the mid 1970s the efforts of national and local governments and international development partners and funders have mainly focused on the establishment of new services. This has been understandable for at least three reasons. First, there has been a deep awareness of the numbers of people still lacking access to an adequate water supply; the water supply problem has been framed in terms of the gap between those 'served' and those still 'unserved'. Awareness of this gap provided the impetus for the first United Nations International Drinking Water and Sanitation Decade, 1981–90 (UN, 1977). It also framed the Millennium Development Goal (MDG) targets in water and sanitation (UN, 2000). The same emphasis remains true at the time of writing, and the access gap is indeed one component of the bigger problem of rural water supply. However, as I argue below, a myopic focus on 'access' can blind us to the bigger picture of sustainable water provision.

Second, it is a reality that the initial construction of new water sources (or for that matter roads, power supplies, schools, or clinics) seems to be more exciting and seductive than the apparently more humdrum work of keeping services working. This aspect of human psychology seems to be replicated across many sectors of economic and development activity, from health and education to infrastructure and public works, and – notably in the context of this book – in the mere provision of 'taps and toilets' rather than actually bringing about the habitual use of satisfactory water and sanitation services and hygiene practices. I argue that valuing the new over the routine, the spectacular over the everyday, gets our priorities wrong if we truly want to see everyone enjoying safe and sustainable water supply services.

Third, rural community water supply interventions are undertaken by various partnerships of governments, their international partners and funders, and national non-governmental organizations (NGOs). All of these have limited budgets, and even more limited budgeting time horizons. None feel capable of accepting long-term responsibility for the services that they initiated. All are keen to devolve responsibility for post-construction management and financing to someone else – in the case of rural water, most often this being the 'community'. One unintended consequence of international access targets such as those enshrined in the 1981–90 Water Decade, the MDGs, and now the Sustainable Development Goals (SDGs), is that insufficient emphasis is placed on keeping the water flowing. Consequently, this chapter and Chapter 8 focus specifically on this dimension of the problem.

Answers to a number of key questions help to categorize different approaches to rural water supply. These questions include:

- Whose initiative was it to establish the water supply?
- Who paid the initial capital cost of getting the service going?
- Who has responsibility for managing the supply after construction or upgrading?
- Who pays the recurrent costs of repairs, extensions, and future upgrades?
- How, and to whom, is the part of the total costs covered by water users actually paid?

Bringing about an enhancement to the water supply system may be at the initiative of a single household or community. It may be at the initiative of government with or without assistance from external funders. Or it may be undertaken by national or international non-government and faith-based organizations. On occasions these organizations work independently of local and national government. I and many other development professionals believe operating independently of government is detrimental because it undermines the mandate of national and local governments, while failing to address what such external organizations perceive as government ineffectiveness.

The management of a rural water supply service post-construction may be the responsibility of a single entity, or the joint responsibility of several organizations. Primary (but not necessarily sole) responsibility may lie with the community, with local government, or with private (either for-profit or not-for-profit, but in reality not profit-making) operators, which may include NGOs, social enterprises, and faith-based organizations.

This chapter explores the range of approaches to 'getting the water going' and 'keeping it flowing', reflecting on the realities and challenges of these various strategies. None is perfect, and no single approach is universally applicable. All approaches have a part to play in improving the lives of rural people in low-income settings. In its own planning, each national government has to determine the best mix of approaches, and how they should evolve over time.

Establishing rural water services

Self-supply

In the absence of external interventions, individuals, households, and communities take their own initiatives to find water and make it accessible to themselves and usually to the wider neighbourhood. In favourable climates, households harvest rainwater from rooves and ground surfaces (RWSN, n.d., c). Where water is collected from swamps and open water bodies, it is common for communities to undertake simple measures to enhance physical access. Naturally occurring springs are often protected in a rudimentary manner, and spring flows are sometimes directed by channels or pipes closer to where people live. Individuals and households construct hand-dug wells, or pay local well-diggers to do so (RWSN, n.d., a); they contract the services of manual well drillers where these exist (RWSN, n.d., e); and in some cases wealthier individuals avail themselves of the services of mechanical well-drilling contractors (RWSN, n.d., f). Some households undertake treatment of their drinking water, using various kinds of filtration technology in the home (RWSN, n.d., d).

All these self-help initiatives to enhance access to rainwater, surface water, and groundwater, and to improve water quality, are collectively referred to as self-supply. They are the subject of Sutton and Butterworth's (2021) book on the topic, which provides a comprehensive global review of such initiatives, and which sets out some of the ways in which governments and others can support and integrate them into a plurality of national approaches to rural water supply.

A number of key observations about self-supply are pertinent to this book. First, self-supply is found everywhere. It is referred to using different terms – including private (DWI, n.d.), informal (Liddle et al., 2014), and off-grid – but choice or necessity means that self-supply is an appropriate solution for some people, in some places, in all countries. It is one element of the national mix of approaches to rural water supply.

Self-help initiatives often emphasize convenient access to water. Domestic rainwater (roof-water) harvesting and household wells and boreholes exemplify this priority for water users. As long as aesthetic water quality – appearance, taste, and smell – is perceived to be acceptable, consumers often place less emphasis on microbiological and chemical water quality. This generalization may be reversed in times of epidemic cholera or other diarrhoeal diseases, but evidence suggests that, in more normal circumstances, compliance with efforts to improve water quality (i.e. correct, consistent, and continual use) can be low (Reygadas et al., 2018).

Consumers' wish to enjoy proximity of access is generally in line with at least one aspect of the SDG goal of 'safely managed' water supply. A safely managed service is defined as one that is accessible on the premises, available at least 12 hours out of every 24, and meeting national or World Health Organization standards of water quality (WHO, 2017b). A self-supply service delivered via a household well or a roof-water collection system comes very close to this component of the definition. If such proximity also allows for water consumption at levels significantly above survival rations, then it is likely that personal and home hygiene will also improve (Thompson et al., 2001).

Self-supply also establishes ownership firmly in the hands of those who initiated the supply. It is at least plausible that this in turn increases the likelihood that the water service will continue to be cared for and managed well into the future.

There is considerable potential to extend and support self-supply through carefully targeted provision of advice, incentives, and subsidies. Where this has been done, many more people have consequently experienced enhanced water supply services (WSP, 2002).

Despite all these observations, governments, international funding organizations, and NGOs share some reservations about the place and value of self-supply. First, there is a general concern that the quality of water provided by self-supply may fall below national or WHO standards, especially in regard to faecal indicator organisms. Governments and water professionals are understandably cautious about promoting an approach to water supply that they believe may lead to sub-standard services. However, this caution should be set in the context of the general water quality of untreated sources; whether such sources are constructed by self-supply or through externally initiated community water supply programmes, some water points will show non-zero thermotolerant coliform counts. Furthermore, untreated water from all point sources 'is likely to undergo further deterioration in quality during transport or storage before drinking' (WHO, 1984: 22).

Second, in the era of the human right to water, some international organizations are unsure about where self-supply fits into the 'rights-holder'/'duty bearer' dichotomy that the rights literature uses. Many, if not all, of the other economic, social, and cultural rights (CESR, n.d.), such as the rights to food, work, and housing, require a high degree of self-help for their realization. As pointed out by Sutton and Butterworth (2021), the obligations of national governments to *respect* and *protect* economic, social, and cultural rights provide a strong argument for self-supply as the natural first priority, while the obligation to *fulfil* the needs of their citizens is relevant in those cases in which households cannot effectively provide for themselves. This book is mainly concerned with the latter situation, which I argue is widespread. Self-supply fills an important niche, but in order for most rural people to experience steady improvements in their water services, external initiatives are needed, with subsidies – and at scale.

Externally initiated supply

The achievement of a situation in which all people, urban and rural, wealthy and poor, enjoy continuing access to sufficient safe water has long been the ambition of national governments and international organizations. The first UN Water Decade (1981–90) had as its goal 'to provide all people with water of safe quality and adequate quantity and basic sanitary facilities by 1990, according priority to the poor and less privileged and to water scarce areas' (UN, 1977: 65). In setting this goal, the international community recognized the right and the need of all people for sufficient safe water, and the imperative of international cooperation to achieve that end (Box 7.1).

A global determination to extend access to drinking water to the entire global population, and to progressively enhance the reliability and quality of the supply, has characterized internationally agreed target-setting since the 1980s, including the present-day SDGs.

Box 7.1 Extract from the 'Mar del Plata Action Plan' in relation to community water supply

a. All peoples, whatever their stage of development and their social and economic conditions, have the right to have access to drinking water in quantities and of a quality equal to their basic needs.
b. It is universally recognized that the availability to man of that resource is essential both for life and his full development, both as an individual and as an integral part of society.
c. To a significant extent similar considerations apply to all that concerns the disposal of waste water, including sewage, industrial and agricultural wastes and other harmful sources, which are the main task of the public sanitation systems of each country.
d. The fundamental challenge facing all mankind can be met only with full international co-operation in all its aspects, entailing the mobilization of physical, economic and human resources.

Source: UN, 1997: 63–4

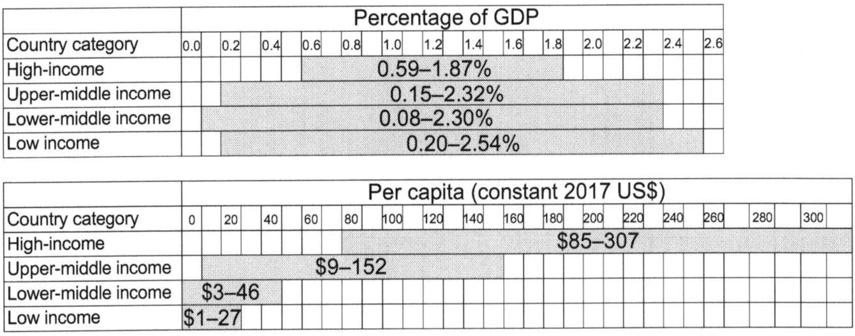

Country category	Percentage of GDP
High-income	0.59–1.87%
Upper-middle income	0.15–2.32%
Lower-middle income	0.08–2.30%
Low income	0.20–2.54%

Country category	Per capita (constant 2017 US$)
High-income	$85–307
Upper-middle income	$9–152
Lower-middle income	$3–46
Low income	$1–27

Figure 7.1 WASH expenditure as a percentage of GDP and per capita for 35 countries
Source: Adapted from Figure 3.12 of UN-Water, 2019: 37

Most of the low- and lower-middle income countries, which are the main focus of this book, pursue their national water supply access targets using a combination of internally generated tax revenues and externally derived transfers of foreign aid. Some countries' water sectors are extraordinarily aid-dependent, a situation that is unhealthy for several reasons – not least because it would suggest limited political will to see beneficial change.

The poorest countries understandably invest the smallest amounts in water, sanitation, and hygiene (WASH) on a per-person basis, although the investment as a percentage of GDP is less variable (Figure 7.1). It is notable that the high-income countries invest significantly more than the middle and lower-income countries, despite the fact that they enjoy high (near-100 per cent) levels of coverage – continual spending is needed to keep the water flowing.

A key finding of successive UN-Water Global Analysis and Assessment of Sanitation and Drinking-water (GLAAS) reports is the high proportion of overall WASH spending coming from households (66 per cent in the 2019 report), compared to governments themselves (22 per cent), loan finance (9 per cent), and external sources (3 per cent). These proportions are, however, highly variable between countries.

A general observation in regard to public financing of WASH in general, and rural water supply in particular, is that only a very limited proportion of the national budget trickles down to local governments. While many countries have decentralization policies, these frequently fail to deliver meaningful fiscal decentralization. In other words, local governments have the *responsibility* for rural water services, but they lack the *resources* to fulfil that duty.

An important milestone in defining how internationally agreed development targets should be pursued at national level came with the Paris Declaration on Aid Effectiveness (OECD, 2005). This was followed by the Accra Agenda for Action (OECD, 2008), the high-level meeting on aid effectiveness in Busan, South Korea, in 2011, and subsequent monitoring of the Paris Principles. The Paris Principles themselves are summarized in Box 7.2.

> **Box 7.2 The Paris Principles on aid effectiveness**
>
> **Ownership:** Developing countries set their own strategies for poverty reduction, improve their institutions, and tackle corruption.
>
> **Alignment:** Donor countries align behind these objectives and use local systems.
>
> **Harmonization:** Donor countries coordinate, simplify procedures, and share information to avoid duplication.
>
> **Results:** Developing countries and donors shift focus to development results and results get measured.
>
> **Mutual accountability:** Donors and partners are accountable for development results.
>
> *Source:* OECD, 2005

The entire thrust of the Principles – which have been widely agreed by national governments and their international donors – is that national governments should be in charge of their own development. Foreign donors and, by implication, international NGOs, should not act in ways that undermine this national sovereignty. Donors that exert undue influence over national governments, and international non-governmental organizations (INGOs) that operate independently of national and local governments, act in defiance of these important principles. The sad reality is that both these abuses of power, exacerbated by the asymmetry of power between those with funds and those needing them, are commonplace.

Extending and enhancing rural community water supply in lower-income countries is an activity shared by national and local governments and NGOs. The relative magnitude of each partner's contribution varies from country to country. However, in the poorest countries three generalizations can be made: first, the majority of the small rural water budget at national level tends to be spent on personnel and general administrative overheads; second, local governments are starved of financial resources and sufficient experienced staff, so rendering their contribution rather limited; and third, international aid organizations, including UN agencies such as UNICEF and other INGOs, wield disproportionate power and influence, well-meaning though that influence undoubtedly is.

In concluding this section, three key points are important for those who are working to enhance rural water services in lower-income countries:

1. For real progress to be made, national governments will need to devote significantly greater proportions of national budgets to water supply in general, and rural water supply in particular (given the high proportion of rural inhabitants, and the higher levels of poverty in rural areas).
2. Far greater proportions of national water supply budgets should be devolved to local governments, together with the personnel and resources (such as vehicles) to enable these organizations to develop their capabilities and effectively invest in and support rural water supply.

3. International donors and NGOs should always work with, and wherever possible through, governments. Where government commitment is unsatisfactory, policies are inappropriate, or capacities inadequate, then international organizations should work with national civil society organizations to influence change. However, in doing so they should scrupulously respect national sovereignty and self-determination, and work hard to mitigate dependency.

Managing rural water services

It is somewhat artificial to discuss management of water supply separately from its financing, and in Chapter 8 the two topics are linked. First, however, it is important to set out the various functions of management, and how these are organized in specific cases.

It is helpful to separate the *attributes and functions* that are required of the water management arrangements from the *organizational set-up* for expressing these. Whaley and Cleaver (2017) have suggested that eight such attributes are necessary in the context of community-managed rural water supply (Box 7.3). With minor modifications, these may also be appropriate for services that are directly managed by public authorities or, with government approval, by private operators.

The assumption behind this list is that certain attributes and capabilities are needed, whichever entity has primary responsibility for management. That entity must have competent leadership that is clear about its (and others') roles, appropriately connected, and recognized both by water users and public authorities (attributes 1, 2, 6, 7, and 8). It must acquire and manage funds, deploying them in maintenance and repair work (attributes 3 and 4). And in so doing it must ensure that all users have access to water (attribute 5).

Box 7.3 Eight attributes required of rural community water management arrangements

1. Authoritative leadership exists
2. Has the capacity to make and enforce decisions, including on rules-in-use*
3. Collects or sources, manages, and accounts for funds
4. Undertakes and/or secures maintenance work
5. Represents all users in a way that ensures equitable access to water supply
6. Recognized as legitimate by both users and the local governance structure
7. Is aware of its roles and responsibilities and the roles and responsibilities of others
8. Is meaningfully linked to other relevant stakeholders

* According to Ostrom, 'Rules-in-use ... are the dos and don'ts that one learns on the ground that may not exist in any written document. In some instances, they may actually be contrary to the dos and don'ts written in formal documents' (2007: 23). Rules-in-use are contrasted with rules-in-form.

Source: Whaley and Cleaver, 2017

All these attributes are necessary, but I would argue that four aspects are crucial. First, motivations and incentives matter. The entity responsible for managing the water supply service, and the users of that service, must be strongly motivated to keep it working (and repair it when it breaks down). Without a strong drive to keep the water flowing, the needed tasks will be neglected, and ultimately the supply will fail. The water consumer aspects of this are discussed further in Chapter 10.

Second, the quality of decisions made and actions taken matter. The management entity must be competent to make good decisions about maintenance and repair, and able to deploy or source the necessary knowledge, skills, and equipment to effect repairs. Professionalism, in the basic sense of competence and effectiveness, is needed, whether or not those exhibiting it are paid.

Third, the sufficiency of recurrent funding is fundamental. Not only must funds be raised for repairs, but those funds need to be sufficient in magnitude to enable good quality work, using good quality replacement parts, tools, and materials.

And, fourth, help must be available. The entity responsible for management must have recourse to others with specific skills, know-how, or authority, if events occur that it cannot deal with alone.

Who should manage rural water supply?

Keeping the water flowing is the primary responsibility of households in the case of self-supply. In the case of externally initiated community water supply, it is the responsibility of communities. The public sector (in the form of a municipality or utility) may have primary responsibility, especially in situations where the management of urban water services can extend outward to the rural hinterland. In some cases it is the responsibility of so-called private operators (which may be for-profit or non-profit organizations) where multiple piped systems and/or water points can be grouped and managed together.

Several reports have categorized the ways in which rural water services are managed and maintained in low- and middle-income countries. The World Bank (2017a) identified five distinct management models (community-based management, direct local government provision, public utility provision, private sector engagement, and supported self-supply) with 16 variants. WaterAid identified 10 configurations excluding self-supply (Lockwood et al., 2018). RWSN (2019) identified six main types of management model, with 14 variants. Lockwood (2019) identified three broad approaches to maintenance within community-based management models (ad hoc reactive, structured proactive, and guaranteed service approach).

Self-supply services are in many respects the easiest to manage, because ownership and management responsibilities are clear, the number of users is small, and technologies generally match the competence of those responsible for their management.

Community-managed services are undoubtedly the most challenging of all the alternatives, since they rely heavily on the voluntary or semi-voluntary work of community members (who may not necessarily be recognized leaders), who may find it difficult to extract user fees from their peers, or settle conflicts among them. In many cases, too, those responsible for management are un- or underqualified to do so, and they may have received little or no training for the task. Unsurprisingly, therefore, the community management model has become the focus of criticism – some constructive, and some much harsher and more negative. I argue below that community management still has an important place and, because of its continuing importance, I devote most space to this model.

Services that are managed by public sector entities (local governments and municipalities), but where water operations are not ring-fenced in budgetary terms, and where specialized institutional capacity is low, tend to perform poorly. In general, it is not realistic to expect such public authorities to perform effectively, and this is not an option that should be encouraged. In contrast, however, the performance of water utilities that reach out beyond their urban centres into the rural hinterlands tend to perform better – although from a financial point of view the rural sector is not particularly commercially attractive to such entities.

In those cases where government delegates responsibility for rural water supply to private operators (including NGOs, faith-based organizations, and social enterprises as well as for-profit organizations), similar observations apply as to well-performing utilities. The combination of a focused mandate, specialized technical and professional staff, and a sound understanding of commercial realities gives such private operators a reasonable chance of success. However, their performance depends crucially on the ability to generate consumer tariffs sufficient to at least cover their direct costs and overheads. Sufficient population density, and the ability and willingness to pay for services are essential pre-requisites for the success of the private operator model.

Community-managed rural water services

Since the 1980s, the dominant model of rural water supply management has been community management, sometimes referred to as community-based management or community-based maintenance. Although maintenance is strictly a sub-component of the broader idea of management, here I use the terms synonymously.

From the earliest days of its roll-out in the UN Water Decade, community management was envisaged as requiring strong internal capacity and strong external support: 'National governments must take specific policy steps to ensure that communities have the capacity, and are empowered, to manage water and sanitation activities ... Today it is recognised that for decentralisation to work and community management to be effective, some governmental entity must provide support on a continuing basis' (Water and Sanitation for Health Project, 1993: 88-9). In a subsequent and more detailed systematic review, this strong linkage between internal (community) capacity and external support was also found to be key to the success of community management (Hutchings et al., 2015).

Many programmes undertaken by governments and NGOs, with or without donor support, that set out to extend access to rural water have constructed water points and systems, yet provided little or no training to communities, and provided no continuing support. This is rather like engineers cutting corners on materials or construction quality, and still expecting their physical infrastructure to work. The fact that two key prerequisites for the success of community management have been so neglected led to the idea of 'community management plus', first articulated by Baumann (2006). The 'plus' refers to the external support that communities need in order to manage their water systems. Had community management been implemented properly, this 'plus' would have been designed into its execution in the first place.

The mediocre performance of community-managed water supplies has been recognized for many years, as evidenced by the following quotes from across half a century of studies on the topic: '(the local authority) constructed the supply scheme and left its maintenance to (the local community). This presupposed a development of civic sense which did not exist, and results have not been satisfactory' (Darling, 1955, quoted in Feachem, 1978: 233); 'experience shows that small community water supplies are often more difficult to be kept running than to construct' (Hofkes, 1981: 32); 'there is increasing evidence that community management has been no more successful in delivering a sustainable water supply than any other approach' (Schouten and Moriarty, 2003: 1).

A plethora of studies of rural water supply sustainability has been undertaken since 2010, all trying to quantify the extent of underperformance and explain its possible causes (Harvey, 2004; Foster, 2013; Marks et al., 2014; Alexander et al., 2015; Chowns, 2015; Fisher et al., 2015; Hope, 2015; Hutchings et al., 2015; Banks and Furey, 2016; Foster and Hope, 2016, 2017; Jimenez-Redal et al., 2017; Whaley and Cleaver, 2017; Foster, Shantz et al., 2018; Foster, Willetts et al., 2018; Foster et al., 2019; Whaley et al., 2019; Olaerts et al., 2019). In quantifying the performance of community-managed water supply, the most commonly used metric has been 'functionality'. At its simplest, functionality is a snapshot observation of whether or not a water collection point is working. When applied to a group of water collection points in a defined area (for example a district or country), or under some form of responsibility of an organization (for example a donor or NGO), it is the ratio (usually expressed as a percentage) of water points that are working to the total number.

Unfortunately, the usefulness of this indicator is flawed, partly because of the difficulty of – and inconsistencies across – organizations in defining precisely what is meant by 'working'. This is compounded by uncertainties over the total number of water points that should be included in the calculation; as well as by the fact that 'functionality' is a measure *at a point in time* (or, more strictly, over the time taken to complete a survey), rather than a measure of performance *over time*. These issues are discussed further by Carter and Ross (2016).

In order to address some of these limitations, work undertaken during the UPGro research programme has proposed a standard method of determining functionality of rural water supply hand pumps (Fallas et al., 2018). This is a helpful initiative, and all attempts to measure functionality in a consistent and clear way, and to introduce measures that result in improved functionality, are to be encouraged. However, functionality still conceals as much as it reveals. A water point that is working today may be broken down tomorrow; one that is broken today may be repaired by tomorrow. More detailed data on times between failures (and their causes) and the time taken to repair (and how this has been achieved) would be of great value.

Community management will continue to be needed in those situations where alternative management models are not appropriate or possible. Self-supply is only a complete solution in situations where the water storage needed for rainwater harvesting is affordable, where shallow good quality groundwater is accessible, or where surface water is abundant and community-level or household water treatment is practised effectively. Management of water supply by public authorities is generally only viable in the form of an extension of geographic responsibility of municipal authorities and utilities. Private operators are only viable where there is a sufficient density of population and water points, and where consumers are able and willing to pay a substantial proportion of the full costs of the service. In all other cases – the majority in many low-income countries – community management will continue, but its performance must improve significantly.

The research studies referred to above have, in essence, asked 'what are the causes of high (or low) functionality, and what is the relative magnitude of their contributions?' They have tried to answer the question through statistical or other quantitative analyses of variables, which have been identified as possible contributing factors. The hidden assumption has been that functionality statistics can be explained by reference to a relatively small set of factors that apply generally. This is somewhat akin to asking 'why do aeroplanes crash?' and hoping for a meaningful generic answer to that question. There are of course several possible contributory causes – including component failure, inadequate maintenance, human error, and so on – but every crash has a unique story, in which one thing led to another, or multiple factors coincided, to cause it.

Just as is the case with an aeroplane, a human body, or a city, many aspects and attributes need to work sufficiently well individually and together if that machine, organism, or socio-economic structure is to function well. In rural water supply, the water resources, the engineering used to gain access to those resources, the technology used to deliver the water, and the way in which that technology is managed and financed all need to work well enough, and work well enough *together*, for the water to keep flowing.

Several writers (Walters and Javernick-Will, 2015; Liddle and Fenner, 2017) have pointed out that rural water services can be conceptualized as a system of nodes and links, with feedback loops; the interactions between different elements in the system make the unravelling of causal linkages (of performance or failure) difficult. Some writers go further and argue that the system is *complex*, exhibiting aspects such as unpredictability and adaptation (Valcourt, Walters et al., 2020). I explore these issues in Chapters 11 and 12. My own view is that it is possible to over-complicate the situation; I believe that we largely understand how to improve the performance of community management, and that if well-established principles were applied rigorously, we would see significant progress. The example in Box 7.4 illustrates this, by reference to a programme that has gone well beyond merely paying lip service to the principles of community management.

Box 7.4 An example of well-performing community management: Kigezi Diocese, Uganda

The Kigezi Diocese Water and Sanitation Programme has been working in south-west Uganda since the mid-1980s. The services provided, largely through piped gravity flow schemes, have achieved a high level of sustainability and continued performance over several decades. Carter and Rwamwanja (2006) concluded that its strong performance was built on taking seriously the well-established principles of community empowerment and support.

The authors argued that the programme was 'doing the right things' by putting into practice the many lessons that had been learned in the sector over the previous decades: full community participation; attention to gender; working closely with other players in local government and NGOs; strong attention to construction quality; being explicit about the need for ongoing support; energetically promoting hygiene and sanitation; building the capacity and self-esteem of communities; and planning strategically for optimum cost-effectiveness.

They argued that 'doing things right means attention to **process** as well as activity. It is a question of **how things are done**, not just what is done. The way staff are treated within the programme (with respect and compassion, giving them a voice regardless of job function or seniority, and with openness) and the way communities and public authorities are treated by the programme (also with respect, a spirit of partnership, and a willingness to learn) determines the effectiveness of the implementing agency both as an entity and in its dealings with the communities where it works. The quality of leadership, and the qualities of the implementing agency, with particular emphasis on transparency and willingness to learn, fundamentally affect its effectiveness.'

Finally, they argued that a key to success was 'doing things for the right reasons': 'there is a deeper level ... the motivation, values and ethos of the implementing agency. If "doing development" is simply a job, a means of paying the bills, then it is unlikely that the process factors touched on under "doing things right" will remain in place for long. The work is then reduced to a matter of stacking the building blocks with no particular commitment to the process or the outcome. If on the other hand there is a strong passion for the work, driven by compassion for the marginalised, driven by humanitarian or religious motives, or motivated by strong instincts for social justice and equity, then the rest can follow.'

Source: Carter and Rwamwanja, 2006: 26

118 RURAL COMMUNITY WATER SUPPLY

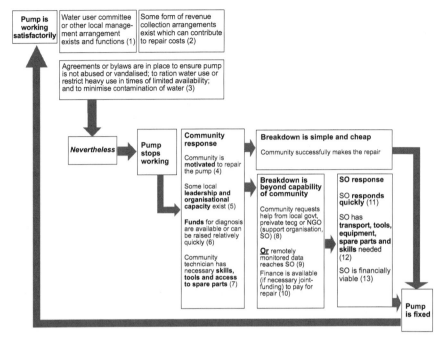

Figure 7.2 Necessary actions before and in response to handpump breakdown
Source: Carter, 2019a

In the case of rural handpump services, a simple diagram (Figure 7.2) illustrates the responses and actions that need to take place when (not if) a water point stops working. The 13 components of this framework can form a checklist to test the adequacy of community water supply management arrangements. In the figure, components (1) – (3) are pre-requisites, which the community can put in place and keep in place while the water point is working. Components (4) – (7) outline the community's own response to breakdown, building on their motivation to keep the water flowing. Components (8) – (13) see the community reaching out for and receiving assistance in those cases in which the breakdown is beyond their own capacity.

This framework and checklist were applied in an evaluation of a rural water programme in Malawi in 2020, which is achieving a high degree of success in sustainable community-based maintenance (Box 7.5).

In the commonly occurring cases in which there are significant weaknesses in components (4) – (13), the development of community capability and the response of support organizations, be they public or private sector, is not necessarily easy. Understanding and addressing the underlying causes of such weaknesses forms an important first step in the implementation of systems strengthening approaches (see Chapters 11 and 12).

> **Box 7.5 Community-managed handpumps in Malawi: the Madzi Alipo programme**
>
> The Madzi Alipo programme has been operating in Blantyre District since 2013. Its sole aim is to strengthen and support communities to manage and maintain their existing handpumps. The programme works in one administrative unit (Traditional Authority – TA) at a time, taking one to two years to undertake baseline functionality surveys, provide targeted training to water user committees (WUCs), and enable those WUCs to repair broken pumps. Functionality rates across TAs by the end of the intervention are typically raised to between 85% and 90%.
>
> The programme's public online database of waterpoints (https://www.madzialipoapp.org) is a key aspect of its ability to keep track of water point performance. The database is 'fed' during field visits by the programme staff using a smartphone app, and also through calls made to WUC personnel by the programme's telephone call centre. In 2020 the programme started installing flow sensors on handpumps, and the data and warnings provided by these sensors greatly increase the effectiveness and efficiency of both monitoring and response.
>
> Over the years following the initial intervention, the programme provides support of various forms – refresher training, spare parts quality control, parts loans, and technical trouble-shooting. The annual cost of this external continuing support is estimated to be about 20% of the initial intervention cost. The need for such continuing external subsidy is unlikely to diminish in the short-term.

Utility management

It is increasingly common for public sector water utilities that are responsible for urban water services to extend their remit to the surrounding rural populations. Where population densities are relatively high, water resources are sufficient, and where rural water consumers can afford the water tariffs involved, this is a natural development of piped supply to increasing numbers of people. It is a welcome development, as long as rural water users do, as a consequence, actually enjoy a reliable and affordable supply. The challenges (for the utility) of billing, revenue collection, and monitoring, as found by the World Bank (2017a) are, however, significant.

Private operator management

Private operator management has been gradually rolling out since around the turn of the millennium (Kleemeier and Narkevic, 2010). Although its spread has been relatively slow, there is still strong interest in the model, where it can be viable. The considerable logistical costs of managing multiple rural water points and systems that are spread out over a large geographical area somehow need to be covered through a combination of water user tariffs and other funding sources. The main challenge, even assuming that national policies allow or encourage the place of private operators, is the identification of sources of steady funding other than tariffs charged to water users, who generally cannot or will not cover the full costs alone.

Five interesting and varied private operator models were identified and researched in the Uptime initiative (McNicholl et al., 2019). All show the ability of this management model to lead to high levels of functionality and short downtimes of water points, but none is fully financially viable in the long term. External subsidy is needed, as with urban water supply and a wide range of other public services.

Conclusions

In those low- and lower-middle income countries with high proportions of their populations living in rural areas, management of water supply will have to be undertaken by a mix of self-supply (with or without external support), community management, extension of utility services, and private operator management. In many countries, the relatively limited opportunities for self-supply, and the slow rollout of utility- and private operator-provided services, mean that community management is here to stay for several years still.

It remains my view that national and local governments and their development partners must strive to make community management more effective – through provision of action-oriented monitoring, financial and technical assistance, conflict resolution, and training. When community-based management is implemented and supported well, it can work. When lip service is paid to principles of community engagement, and reliance is placed on short, one-off training sessions, it is hardly surprising if community management fails.

Keeping the water flowing cannot be done on the cheap. National and local governments need to set aside sufficient resources to deploy the technical and professional staff, together with vehicles and equipment, which are needed to complement communities' own skills and resources. Community management is not, and never should have been, a matter of external organizations constructing water systems, handing them over to communities, and walking away. Until this important lesson is taken seriously, the sustainability challenge will remain.

CHAPTER 8
Finance: the fuel for sustainable rural water services

Abstract: *This chapter provides an overview of the financial requirements necessary to establish and manage water, sanitation, and hygiene systems. The life-cycle costing procedure distinguishes the capital costs of first-time access from the minor operation and maintenance costs, major capital maintenance expenditures, the direct and indirect support provided by local and national government, and the costs of capital. Capital costs may represent less than half the total cost of the service, with the majority required to keep services working. It may also cost four times or more to provide and maintain safely managed services compared to basic services. The gap between available and required funds is large and lower income countries may need to spend as much as 5–10 per cent of annual GDP to meet national targets. The full life-cycle costs of water provision are also often unaffordable at the household level. Benefit/cost ratios of access to improved water reveal that the societal benefit of investment in improved drinking water generally exceeds the costs involved. There are, however, other reasons than the narrowly economic for extending services to all. Rural water programmes are not financially viable if reliance is placed only on user tariffs to finance the full (life-cycle) costs. Urban water supply, and many other public services, are subsidized from taxes, but increasingly there needs to be a willingness by governments and donors to contribute to the recurrent costs of rural water services.*

Keywords: life-cycle costing, affordability, economic returns, benefit/cost ratio, financial viability, subsidies

> 'Money is like muck, not good unless it be spread'
> —Francis Bacon, 1601

Introduction

Significant sums of money are needed to provide first-time access to the physical infrastructure that can provide a good level of service to its users. Even more is needed to keep services working and to continue to move people up the ladder of water service levels (see Chapter 2). Counting the cost is the first step national governments and their partners need to take as they invest in rural water supply.

Important questions arise as to what level of investment – and consequently what level of service – national governments, other implementing agencies, and water users themselves can realistically afford. There may be a danger of one-size-fits-all international goals and targets – especially the Sustainable Development Goal (SDG) 6 'safely managed water' target – exacerbating inequalities, which are already rife in lower-income countries.

Investing in rural water supply has potential economic benefits to society, and careful analyses can demonstrate its economic value. It is, however, important to note the limitations of economic benefit/cost analyses. The wider question of how investment in rural water supply should be justified is picked up in Chapter 12. Suffice to say here that economic arguments may be part of the story, but they do not paint the whole picture.

The analysis of future costs is necessary to understand the likely financial viability of services, whether they are provided by individual community-managed water points or by multiple water delivery systems managed by private operators or public utilities. When revenues – from tariffs, taxes, or transfers – fail to keep up with the necessary costs of keeping services working, they inevitably fall into decline.

The recurrent costs of rural water services can only ultimately come from the users of those services (in the form of tariffs), from the state (originating in taxes), or from foreign donors and lenders (as transfers of official development assistance). It is increasingly clear that rural water users cannot shoulder the entire burden of cost; as in other areas of public services, it should increasingly become the norm for the state to share that liability.

By thoroughly analysing the costs, affordability, wider benefits, and financial viability of rural water supply, two simple conclusions emerge. First, nation states (with or without the assistance of foreign donors and lenders) should invest more – much more – in rural water supply. And second, the balance of that investment should progressively shift from capital outlay on new systems to the recurrent expenditures needed to keep services working well.

Counting the cost

The WASHCost project (2008–13) set out to understand the costs of water, sanitation, and hygiene services. The project was funded by the Bill and Melinda Gates Foundation, implemented by IRC (Netherlands), and its detailed investigations were carried out in Andhra Pradesh (India), Burkina Faso, Ghana, and Mozambique. Although the project produced time- and place-specific costs of different types and levels of service, its most significant contribution – more so as time passes – was in its structured methodology for analysing costs. The so-called life-cycle costing (LCC) procedure put forward by WASHCost is now used widely as the industry standard for understanding the costs of water and sanitation services internationally.

Life-cycle costing was first developed in the United States in the mid-1960s by the Department of Defense (Estevan and Schaefer, 2017). It emerged in recognition of the fact that the cost of acquisition of weapons systems represented only a small part of the total costs, which also included operation and support costs and ultimate disposal. LCC assumes that a product, service, or works has a finite life. Public procurement in the European Union requires the use of LCC to cover:

> parts or all of the following costs over the life cycle ... (a) costs, borne by the contracting authority or other users, such as (i) costs relating to acquisition, (ii) costs of use, such as consumption of energy and other resources, (iii) maintenance costs, (iv) end of life costs, such as collection and recycling costs; (b) costs imputed to environmental externalities linked to the product, service or works during its life cycle, provided their monetary value can be determined and verified; such costs may include the cost of emissions of greenhouse gases and of other pollutant emissions and other climate change mitigation costs. (European Union, 2014, Article 68)

In rural water supply, although all individual physical assets or components of assets have expected lifespans, the service itself does not. Many physical assets (e.g. handpumps, pipelines) can effectively continue to provide service indefinitely, if failed components are replaced as and when necessary. Others (e.g. reservoirs, boreholes) eventually come to the end of their useful life and have to be replaced in their entirety.

Although the term life-cycle costing might suggest a limited time horizon, this is not the intention of those organizations that use the approach for costing of water supply and sanitation services. IRC for example define life-cycle costs as:

> the costs of ensuring adequate services to a specific population in a determined geographical area – *not just for a few years but indefinitely*. All costs from construction and installation, to maintenance, repairs and eventual replacement are taken into account, including payment for borrowed money either at household or national level government. Life-cycle costs also include costs for source protection, training and capacity development, planning and institutional pro-poor support. In short: the costs that it takes to deliver a service and not only to build infrastructure. (van der Kerk, 2015, emphasis added)

This emphasis on indefinite service echoes that of WaterAid's Sustainability Framework (2011): 'Sustainability is about whether or not WASH services and good hygiene practices continue to work and deliver benefits over time. No time limit is set on those continued services, behaviour changes and outcomes. In other words, sustainability is about permanent beneficial change in WASH services and hygiene practices.'

The WASHCost project published benchmark cost ranges of different categories of point-source and piped water supply systems, using the categories in Table 8.1 (WASHCost, 2012). Since these 2011 cost benchmarks are now increasingly outdated, I summarize them here (Figure 8.1), not as absolute dollar costs, but rather as percentages under each cost category. Capital costs (capex) are recalculated as equivalent annual costs of the initial investment,

Table 8.1 Life-cycle costing: main cost components

Cost components		Summary description
Capital expenditure The costs of providing a service where there was none before; or of substantially increasing the scale or level of services	Capital expenditure hardware (capex)	Capital invested in constructing or purchasing fixed assets such as concrete structures, pumps, and pipes to develop or extend a service
	Capital expenditure software (capex)	The costs of one-off work with stakeholders prior to construction or implementation, extension, enhancement, and augmentation (including costs of one-off capacity building activities)
Recurrent expenditure Service maintenance expenditure associated with sustaining an existing service at its intended level	Operational expenditure (opex)	Operating and minor maintenance expenditure: typically comprises regular expenditure such as labour, fuel, chemicals, materials, and purchases of any bulk water
	Capital maintenance expenditure (capmanex)	Asset renewal and replacement cost: occasional and 'lumpy' costs that seek to restore the functionality of a system, such as replacing pump rods or foot valves in handpumps, or a diesel generator in motorized systems
	Cost of capital (CoC)	Cost of interest payments on micro-finance and loans used to finance capital expenditure. Cost of any returns to shareholders by small-scale private providers
	Expenditure on direct support (expDS)	Expenditure on support activities for service providers, users, or user groups
	Expenditure on indirect support (expIDS)	Expenditure on macro-level support, including planning and policy-making to decentralized district, municipal, or local government

Source: Burr and Fonseca, 2013

applying commonly used factors (a discount rate of 10 per cent and a lifespan of 20 years). The use of a 5 per cent discount rate alters the relative proportions of capital and recurrent costs (reducing the equivalent annual cost of capex from about 50 per cent to 40 per cent, and increasing the recurrent cost proportion to about 60 per cent), but the overall patterns remain unchanged. Cost of capital and expenditure on indirect support (expIDS) could not be included in WASHCost's analyses, because of inadequacy of information on these components.

It is notable in Figure 8.1 that, for all types and sizes of water supply system, the capital cost only reflects about half the total cost. Minor operation and maintenance costs are approximately 10 per cent, while major replacement costs (capmanex) lie in the range 20–25 per cent. Support costs (whether by

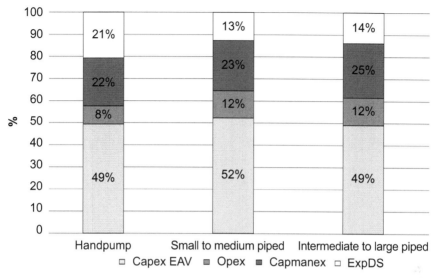

Figure 8.1 Cost proportions for different cost components
Source: WASHCost, 2012, recalculated for plotting

local government or another entity) range from the low tens to more than 20 per cent.

The WASHCost benchmark costs were deliberately published as cost ranges; it is inevitable that most of these ranges are wide, reflecting important contextual differences between the countries from which the data came. However, it is also important to note that the average benchmark dollar costs published by WASHCost show piped systems (of any size) incurring approximately double the capital and recurrent (opex + capmanex + expDS) costs compared to borehole-handpump systems (Figure 8.2). Serving people with piped water may be highly desirable, but it comes at a cost.

Several years after WASHCost's report, Hutton and Varughese (2016) carried out an extensive (140-country) and detailed analysis of the costs of meeting SDG targets 6.1 and 6.2. Their analysis has important implications for affordability, and we will return to this matter later. However, the subject of interest here is the unit costs assumed by the authors for increasing access to sustained services provided by rural water infrastructure. They quantify three cost components, partially following the WASHCost approach: capex, opex, and capmanex. The definition of capmanex used, however, is 'replacement of parts and renovation or rehabilitation when required *to extend the life of the hardware to its expected life span*' (Hutton and Varughese, 2016: 31, emphasis added). This is a less demanding definition than that used by WASHCost, which envisages capmanex as an integral part of asset management, including 'renewing (replacing, rehabilitating, refurbishing, restoring) assets to ensure that services continue at a similar level of performance as was first delivered' (Fonseca et al., 2013: 11). In a subsequent World Bank analysis using Hutton

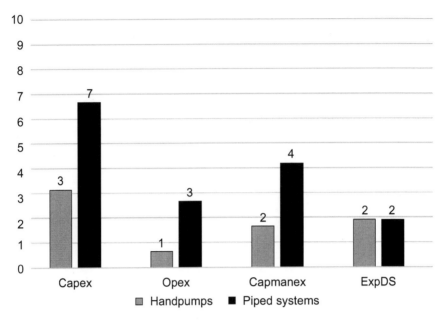

Figure 8.2 WASHCost average costs compared: handpumps vs piped systems
Source: WASHCost, 2012, recalculated for plotting; costs on the y-axis are 2011 US$ per person per year. Capex is recalculated as equivalent annual cost

and Varughese's approach, the authors reintroduce the end-of-life replacement costs of physical assets (Fox et al., 2019).

Hutton and Varughese examined three technology/service levels in their analysis of rural water supply costs: safely managed water; boreholes or tubewells; and hand dug wells. The authors' unit costs and assumptions are set out in Table 8.2. These and the dollar costs shown in Table 8.3 are derived from the spreadsheet model used by Hutton and Varughese (2016).

The summary in Table 8.3 shows that 46–66 per cent of the annualized costs (depending on service level) are recurrent, while the remainder is the

Table 8.2 Summary of unit cost assumptions made by Hutton and Varughese (2016)

Component	Explanation		
Capex	Individual estimates supplied by each country; 5% or 10% added for software		
Opex	Annual cost set at 5% of capex		
Capmanex	Different assumptions according to service level, as follows:	Lifespan (time to capmanex), years	Capmanex as % of initial capex
	Safely managed	20 (10)	30
	Basic household piped	20 (10)	30
	Borehole or tubewell	20 (10)	30
	Dug well	10 (5)	30

Table 8.3 Country-averaged per-person costs in Hutton and Varughese (2016) model

	Capex average ($)	Capex annualized ($/year)	Opex annualized ($/year)	Capmanex annualized ($/year)	Total cost annualized ($/year)
Safely managed	266	21.4	31.3	9.4	62.0
Borehole or tubewell	85	6.8	4.4	3.1	14.3
Hand-dug well	23	3.0	1.1	1.6	5.7

Note: All dollar costs 2015 US$ per person

capital cost. The total annualized costs of safely managed water are more than four times the estimated costs of borehole water services, which in turn are 2.5 times those of water from hand-dug wells.

In short, the conclusions of the various cost analyses outlined above are as follows:

- Whatever the technology and service level involved, typically 50–60% of the real cost is incurred post-construction; operations, maintenance, repairs, and eventual replacement of components and entire assets incur the greater part of the annualized costs. The capital cost is the tip of the iceberg – the smaller and easier part of the investment.
- The higher the service level, the more costly the service; piped (groundwater or surface water) supplies are more costly than point (groundwater) supplies, by a factor of at least two; in the case of safely managed services, the costs are four times as much as the services provided by boreholes and handpumps.

Two further questions are raised by this latter conclusion. In low-income economies, what is the affordable mix of piped services and those provided by point sources? And are the benefits of piped and safely managed services 2–4 times higher than those provided by boreholes and handpumps (or equivalent)?

Affordability of rural water services

Affordability is a much-used term, but clear and simple definitions are elusive. A variety of definitions and indicators of affordability exist, but none is perfect (Hutton, 2012b). In 2019, the authors of the World Water Development Report summarized that 'affordability as a concept will need to be further defined before it can be effectively measured' (UN, 2019: 88). All agree that water should be affordable, but there is no clear consensus on what exactly that means. Here I briefly examine the question of affordability as it applies, first, to national governments and, second, to individual households.

The sums of money that national governments are willing to invest in improvements to, and maintenance of, rural water services depend on numerous factors. Perceptions of the benefits provided by such services combine with governments' awareness of their ability to make budgetary commitments. It is clear that the beliefs and attitudes that underlie willingness to invest, and the actual resources at the disposal of the state – their ability to invest – together determine what is affordable in a real sense. It is also clear that national decisions about budget allocations are as much about politics as scientific or other evidence-based argument (see Chapter 9).

At the national level, the proportion of the state budget allocated to water and sanitation, and specifically to rural water supply, is widely recognized to be insufficient. The 2019 Global Analysis and Assessment of Sanitation and Drinking-Water (GLAAS) Report (UN-Water, 2019) showed the profound inadequacies in national financing of drinking water supply. Fewer than 15 per cent of countries responding to the GLAAS survey reported that funding was sufficient to achieve national targets. The rural water funding gap – the difference between funds needed to reach targets and funds actually available – was largest of all the permutations (rural/urban; water/sanitation) at 78 per cent. In other words, for those countries that could report on this question, budget allocation was only 22 per cent of need.

Hutton and Varughese (2016) estimated the total annualized capital costs to provide universal access to basic water supply and to safely managed water to be US$1.4 billion per year and $13.8 billion per year, respectively. Assuming that the capital cost represents 45 per cent of the total annual cost, then these amounts increase to $3.1 billion and $31 billion. The authors compared figures such as these to the gross product (GP_{140}) of the 140 countries included in their analysis. The total gross product of the 140 countries in 2015 was $29,335 billion. The annual cost therefore of extending access to, and maintaining, rural water services universally would be about 0.01 per cent of GDP for basic services, and 0.1 per cent for safely managed services.

Naturally, totals and ratios such as these hide as much as they reveal. It may be that globally we can afford to get basic, or even safely managed, drinking water services to all by 2030, but the reality is that countries face their individual financial challenges. The world is not 'in it together'. Hutton and Varughese (2016) show that the global average cost of meeting SDG 6.1 and 6.2 (i.e. universal access to safely managed water and sanitation) by 2030 is about 0.39 per cent of GP_{140}. However, their analysis also shows that for lower-income countries, the cost rises rapidly, reaching 5–10 per cent for the lowest-income countries.

Turning to affordability at the household level, most discussions restrict the meaning of affordability to an assumed ability to pay, so leaving out the willingness dimension. This simplifies matters, but there is a concomitant risk of oversimplification. Adopting threshold water and sanitation tariffs, as has been the case in England and Wales, for example, showed that in England 23 per cent of households were spending more than 3 per cent of their income on these services, while 11 per cent were spending more than 5 per cent of

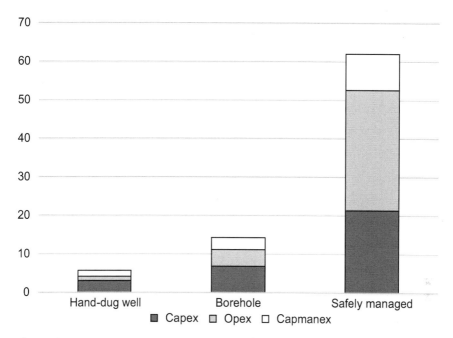

Figure 8.3 Rural water annual costs per person (US$ 2015) by service and cost component
Source: Hutton and Varughese, 2016, cost model

their income (Ofwat, 2015). While water tariffs or charges may be affordable to people on average and higher-than-average incomes, those on very low incomes face real difficulties. In most high-income countries, measures are in place to address such difficulties (Aqua Publica Europea, 2016).

Figure 8.3 presents the cost estimates of Table 8.3 visually. It further reinforces the points made earlier about the relative costs of (a) services provided by hand-dug wells, boreholes, and piped/treated (safely managed) supply, and (b) capex, opex, and capmanex components. In regard to affordability, the total per person annual costs of hand-dug wells, boreholes, and safely managed services represent, respectively, 0.8 per cent, 2.1 per cent, and 8.9 per cent of the widely accepted $1.90 per day absolute poverty level. More than 700 million people in low- and middle-income countries, more than half of whom are likely to be rural dwellers, live in this level of poverty (World Bank, 2020b) – and that number may well rise significantly as a consequence of Covid-19.

The economics of rural water supply

The economics of rural water supply requires some form of evaluation and comparison of the costs and benefits of these services, usually in the form of a benefit/cost analysis. Decisions about spending on rural water services should not be, and are not, taken solely on economic grounds; however, if the economic arguments are persuasive, this can help to unlock investment from ministries of finance and other financiers for whom economic arguments may carry much weight.

Quantifying the costs of rural water supply can be difficult and, as we have seen, published cost estimates often lack comparability. Determining precisely how each cost component should be defined, and establishing how required expenditures should be allocated between cost components, represents just some of the details of the cost side of the calculation. In determining annual equivalent costs, appropriate discount rates have to be selected and decisions taken on whether and how to allow for inflation. However, if costing poses challenges, then quantifying and monetizing benefits is even more of a thorny issue.

When considering the benefits of rural water supply programmes, decisions need to be taken about precisely which benefits to include and which ones to exclude. In an early attempt to quantify the benefits, Churchill (1987: ix) was of the view that 'there is a very tenuous link between improvements in health and investments in water supply and sanitation services. The best that can be said is that these services may be necessary, but not sufficient, to achieve any tangible effects on morbidity and mortality'. Consequently health benefits were omitted from the analysis. Decades later, in contrast, economists take the view that such health benefits are real and quantifiable – although the assumptions that have to be made can still be contentious.

It is common to include time savings on the benefits side of the benefit/cost balance. But bold assumptions still have to be made about how much time will be saved, how saved time will be used, and what that time is worth in monetary terms.

Two key questions arise in economic analyses of rural water services. First, do the benefits of improved water services in general exceed the costs (and, if so, by how much)? And second, are the undoubtedly greater benefits of piped and safely managed water sufficiently large as to offset the significantly higher costs? In addressing questions such as these, we can determine the worthwhileness, from an economic point of view, of investments in rural water. If the costs are high relative to the benefits, so that alternative investments appear more attractive, or if the costs exceed the benefits, then other arguments must be deployed in order to advocate for investments in rural water. Some of these arguments are picked up in Chapter 12.

At the time of writing, the most recent global analysis of the economics of WASH remains that carried out by Hutton (2012a) and published by the World Health Organization. Other analyses had been carried out in the previous 10 years, but here I summarize the pertinent points from Hutton's study.

Hutton's (2012a) analysis was undertaken near the end of the Millennium Development Goal (MDG) period, so unsurprisingly it focused on the end-point of that period, namely the year 2015. It investigated the costs and benefits of reaching the MDG target (reducing by half the gap in access relative to the year 1990), and the much more demanding target of reaching universal coverage by 2015. The study used data from 136 countries. Country-based cost data for rural water borehole access and for piped house connections were assumed, but at the time there was little individual country-level information available, so the same values were used for many countries.

A wide range of potential benefits of improved water and sanitation were identified – 13 in the case of water – but of these only five were actually included in the analysis. The reason for the exclusion of many potential benefits was attributed to lack of evidence. However, some are simply not amenable to simple economic analysis. For instance, time saved that is used for increased leisure or that reduces the burden on women is very difficult to quantify or monetize.

The benefits that were included related to: health (reductions in infectious diarrhoea, malnutrition, respiratory infections, and malaria); savings related to seeking less health care; savings from productive time losses due to disease; and savings related to premature mortality. Time savings were also included, with the assumed time saved (in hours) being valued at 30 per cent of hourly GDP (or 15 per cent for children).

The analyses were conducted by individual country, and then aggregated to regional level. The results of the analysis concluded that investment in improved drinking water with the target of universal access by 2015 would result in an overall economic benefit/cost ratio to society of 2.0. In other words, 'the global economic return on water expenditure is US$ 2.0 per US dollar invested' (Hutton 2012a: 31). However, this global value hides benefit/cost ratios in three of the nine regions that are 1.0 or less; in other words the economic benefits are less than, or at best equal to, the costs (Figure 8.4).

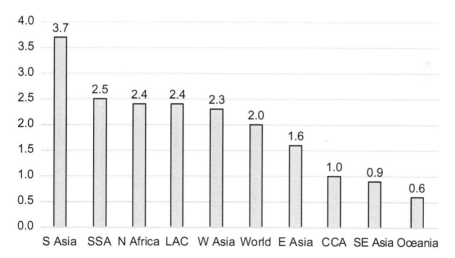

Figure 8.4 Benefit/cost ratios of investments to achieve universal access to improved drinking water by 2015
Note: CCA – Caucasus and Central Asia; E Asia – Eastern Asia; LAC – Latin America and the Caribbean; N Africa – Northern Africa; S Asia – Southern Asia; SE Asia – Southeast Asia; SSA – sub-Saharan Africa; W Asia – Western Asia
Source: Hutton, 2012, re-plotted

The same observation applies to almost a quarter of the countries included in the analysis.

Globally, Hutton found that the economic benefits of sanitation exceeded those of water supply by a factor of nearly nine. Time saving represented 70–90 per cent of all the economic benefit, although in Southern Asia and sub-Saharan Africa the value of saved lives made a major contribution to health benefits.

Finally, I highlight four key conclusions that Hutton drew from his analysis. First, the vast majority of financing of drinking water services is needed for maintaining and replacing existing infrastructure (as opposed to financing new coverage). Second, slow progress in attaining targets tends to increase the future financing burden. Third, focusing on the unserved to the relative neglect of those already served risks under-funding operation and maintenance and consequent slippage of previously 'served' populations to de facto 'unserved' status. And, fourth, achieving universal coverage, even with basic services, is hugely ambitious; Hutton mentions a period of 20–30 years (not the 15 years of the SDG period), and that does not take account of population increases, price increases above the average rate of inflation, and consumers' rising expectations of service levels. Perhaps the 15-year SDG period should more realistically have been framed as a 50–100 year effort (Hutchings and Carter, 2018).

Financial viability of programmes and services

The financial viability of rural water services is fundamental to the sustainability of those services. Put simply, regardless of how rural water programmes, projects, or area-wide services are managed, if the level of income is below that required to operate, maintain, repair, rehabilitate, and extend the service, it will suffer an inevitable deterioration over time to the point of complete failure. Finance is the fuel the service requires in order to keep running.

Revenues for public services such as rural water supply can come from three sources, the so-called 3-Ts (OECD, 2009). These sources are general taxation (which funds national budgets), tariffs (payments made by consumers), and transfers (funds provided by donors through official development assistance).

In rural community water supply it has been the norm for many years that the capital costs – seen as unaffordable by communities – have been paid by governments (with or without donor funding) and NGOs (using international donor funding). The exception to this is self-supply (Sutton and Butterworth, 2021), and the importance of the expenditures made by households has recently been further recognized in the proposal to add a fourth component to the funding mix, namely household investments (Danert and Hutton, 2020).

Beyond self-supply, however, it is ironic that the larger part of the cost (the recurrent expenditures on opex and capmanex) is handed over to the responsibility of the community. In short, those organizations external to the community (implementers and their donors) do the easy bit, while leaving the much more challenging tasks of post-construction management and financing to those who are arguably least equipped –in terms of both skills and resources – to undertake them.

A 2019 review (McNicholl et al., 2019) of the experiences of five large and successful private operators (including non-profit organizations) serving an estimated 1 million people, showed that water users were paying about one quarter of the total costs incurred by the provider. Although the providers were achieving high levels of functionality and low downtimes – in other words very high reliability of service – generally the operational costs were far from being covered by user payments. Moreover, capital maintenance and asset depreciation were not accounted for in the analyses. In short, it is very hard, perhaps impossible, to make services work on the basis solely of tariff payments. One 'T' is not enough.

Rural water services can only be sustainable in the sense used in this book (i.e. indefinite service, having no time limit) if revenues match the necessary expenditures (opex, capmanex, and the direct support costs incurred by the local government or private operator). In low- and lower-middle income countries, significant numbers of people cannot realistically afford to pay the full recurrent costs; they can make a meaningful contribution to these costs, but not cover them in their entirety. This leads us to the final question in this chapter: who should pay what in the endeavour of bringing sustainable water services to all rural people?

Who should pay what?

Assuming that governments, NGOs, and their financing partners continue, gradually, to increase access to rural water services through making the necessary capex investments, the question of 'who should pay for the recurrent costs?' remains.

The general pattern in rural water supply has been that water users and consumers have to cover the full recurrent costs, a self-evidently unsustainable strategy. As insufficient funds are invested in operation, maintenance, repairs, and replacements (opex and capmanex), systems deteriorate to the point where they fail. They then have to await the next round of projectized interventions to rehabilitate or, worse still, reconstruct water assets; either option involves unnecessarily high expenditures.

Franceys et al. (2016) made persuasive arguments for the international donor community to take responsibility for at least part of the recurrent costs of sustaining the assets that they have funded, especially the capital maintenance costs. This responsibility has, to date, not been accepted by the donors, but as long as the national tax base in lower income countries and the ability and/or willingness of water users remain insufficient, there seems to be no alternative.

In their 2020 article, Tutusaus and Schwartz argued that despite international rhetoric and national policies calling for full cost recovery from tariffs, public utilities have to embrace a different reality. They have to 'talk the talk' of full cost recovery, but they simultaneously negotiate grants and subsidies from the public purse to make ends meet. The authors (rather unhelpfully, in my view) refer to this as 'organized hypocrisy'; perhaps realpolitik would be more accurate. And maybe financing organizations and governments should critically review the myth of full cost recovery.

The point, however, is that urban utilities and nearly all other public services are subsidized by governments, often with a proportion originating from donor funds. Why, other than its invisibility, is the rural water supply sector generally perceived to be an exception to this rule?

It is one thing to call for subsidies, but quite another to ensure that they target and reach the right people without causing market distortions or resulting in capture by those who do not need them. Another study of subsidies to users of piped water services in 10 countries in Africa, Asia, and Latin America showed that 'richer households enjoy a greater share of subsidies since the poor have less access to piped water and, even when connected, consume less water generally' (Abramovsky et al., 2020: 12). They conclude that in order to improve the targeting of subsidies, the main measure should be to improve households' access to services – echoing Franceys' spotlighting of the absurdity of charging households which are in dollar-a-day poverty an average of $295 simply to get connected (Franceys 2005a, 2005b).

CHAPTER 9
Rural water users and community water supply programmes

Abstract: *Rural communities in low- and middle-income countries exhibit high levels of income poverty and multi-dimensional poverty. It is estimated that 80 per cent of those lacking safe water live in rural areas; 80 per cent of the extreme poor live in rural areas; and 74 per cent of rural people in low-income countries suffer multi-dimensional poverty. Divisions and inequalities are widespread in rural communities; exclusion and vulnerability are common. Furthermore, rural communities often experience political, demographic, and environmental pressures. Much rural water programming has failed to take account of this, instead applying a one-size-fits-all approach to community engagement. Crucially, many programmes have failed to apply good practices, which have been learned over many decades by the best organizations, resulting in significant negative impacts on individuals, households, and whole communities. Free, meaningful, and inclusive participation of communities is essential for sustainable water services. This requires those implementing community water programmes to understand how individual communities work, involve community leaders and women, gain official support, and make special arrangements to include groups that might otherwise be excluded.*

Keywords: rural communities, poverty, multi-dimensional poverty, exclusion, gender, rural water programmes, participation, equality, inclusion

> 'We need boreholes because we rely on unsafe water from streams and unprotected wells. It is a critical problem because most of these streams and wells dry out during the dry season'
> —A rural water user, Malawi, quoted in Narayan et al., 2000: 73

Introduction

The focus of this book thus far has been more on the technical and infrastructural elements of water supply for rural households and communities in low- and middle-income countries. In this chapter I first turn greater attention to rural people – the users of domestic water in rural areas. What are the attributes of rural water users and rural communities in low-income settings? What challenges do rural communities face? What strengths and capabilities do rural households and communities bring to the water supply endeavour?

Second, given the nature of rural communities, what can be said about the ways in which rural water programmes have typically engaged with them? To what extent have such programmes – implemented by governments and non-governmental organizations (NGOs), usually with donor support – taken account of rural community realities and applied good practices?

Then, third, I examine the ways in which governments and external agencies could more constructively approach rural communities, in the pursuit of better and sustainable services for all.

Rural communities

Rural living

Rural living describes a wide range of situations. Rural lives may be lived close to trading centres, small towns, and cities, or very remote from such population centres. They may be sedentary or (semi-)nomadic – in dry climates where pastoralism is the only livelihood option, dwelling in one place is a transient experience. Some rural individuals and households are well connected to family members and businesses nearby and in towns and cities, while others have much more limited connectivity with people outside the immediate community.

Rurality can be characterized by distance from conurbations, density of population, and association with arable and/or livestock farming as the main livelihood option. Each of these factors has important implications for rural people. Long distance from markets (often combined with challenging terrain) limits income-generating opportunities; low population densities make provision of health, education, energy, and water services expensive; dependence on agriculture, especially rainfed farming, makes rural people very vulnerable to climate variability and change.

However, as well as the spatial measures of rurality, Okidegbe (2001) reminds us that rural people are principally engaged in agriculture, and their access to land is therefore crucial. He classifies the rural poor into five categories: the landless (those without any crop land); those having a low asset base (smallholder farmers with up to two hectares of cropland); pastoralists (those who are not settled in any specific area and who derive most of their income from pastoral livestock); women-headed households (women who are charged with supporting their families without any outside support); and others (e.g. indigenous populations).

An investigation by Bird et al. examined the poverty of people living in remote rural areas, categorizing these as:

- Areas with 'extreme' ecologies where infrastructure and communication is limited and difficult: mountains, swamps, deserts, islands, chars ...
- Low-potential areas – semi-arid, limited topsoil, water resources; and/or degraded: polluted, saline, landmines

- Poverty 'pockets' where social-political exclusion – on the basis of language, identity (caste, religion, tribe, ethnicity, class, gender) maintain significant proportions of the population in poverty
- Areas experiencing long-term conflict, where the dynamic of conflict itself has led to widespread damage to the resource base as well as people's capabilities. (Bird et al., 2001: 14–15)

The authors are pessimistic about the likelihood of people in such remote rural areas moving out of poverty in the near future. Nevertheless, rural communities have great strengths and resilience – evidenced by their ability to survive and thrive in difficult and sometimes hostile environments – but they are increasingly threatened by internal and external forces that are beyond their control. These threats (including demographic and environmental trends and shocks) exacerbate poverty in general, as well as increasing the challenge of bringing about sustainable water supply for all.

Arguably the biggest threat to rural people, especially those living very remote from cities, is their invisibility. It is too easy for them to be overlooked by decision-makers in government; and given the challenges inherent in serving them, it is more convenient to continue to under-invest in their development.

A note on rural community leadership and organization

Local-level leadership and governance have always been needed to take decisions that affect the entire community, to resolve conflicts within communities, and to make rules about the management of common property such as grazing land and water. Informal customary councils in Afghanistan (Brick, 2008), and traditional village government throughout Africa (for example those described well in Malawi by Hussein and Muriaas, 2007), are simply examples of a more extensive reality. Over the last several generations, 'modern' local government structures have tended to supplant these traditional forms of government, but in many places they still exist and are respected by the communities in which they survive.

As well as these traditional forms of government, various forms of community organization exist, often with specific functions such as to arrange communal work and weddings, and to respond in the event of sickness or death. In many countries, communities organize themselves to undertake savings and loans. Community groups include the funeral associations known as *iddir* (Pankhurst, 2008) and the conflict resolution arrangements known as *gadaa* (Edossa et al., 2005), both in Ethiopia; and traditional savings groups in at least 30 African and Asian countries (Bouman, 1977).

In the implementation of community water supply programmes, it is important to try and understand the existence, nature, and dynamics of change of such local institutions. Although there may be risks of reinforcing existing power imbalances, it may be more effective to build on existing structures and organization than to introduce new forms such as one-size-fits-all water point committees. In any case, it should not be surprising if communities adapt and alter externally introduced structures (Cleaver and Whaley, 2018).

Water, poverty, and disadvantage

The majority of those who lack access to an improved water source ('limited' or 'basic' in Joint Monitoring Programme (JMP) terminology), as well as the majority of those who lack access to a safely managed service, live in rural areas (Table 9.1). Whatever the level of service envisaged, the access gap is felt disproportionately in rural areas.

Poverty more generally is also a predominantly rural phenomenon. Whether poverty is conceived narrowly in terms of income (or consumption), or whether it is recognized as a multi-dimensional experience, poverty is especially marked in rural areas. In their analysis of 'extreme' and 'moderate' poverty in 89 developing countries using the US$1.90 and $3.10 per day thresholds, Castañeda et al. (2018) concluded that 80 per cent of the extreme poor and 76 per cent of the moderately poor live in rural areas. Furthermore, extreme poverty is strongly associated with working in agriculture and children under 15 make up 44 per cent of all extreme poor. However, the evidence regarding gender gaps in poverty is mixed and inconclusive.

Rural life in low-income countries is characterized by a wide range of deprivations, including food insecurity, water insecurity, limited communications, lack of access to energy, few jobs, and limited health and education services. We are accustomed to seeing statistics describing each of these deprivations individually (for example, how many people are food insecure, the extent of unemployment, the numbers lacking access to electricity), but the hard fact is that many rural people experience many or all of these dimensions of poverty simultaneously.

The multi-dimensional poverty index (MDPI) measures deprivation in relation to 10 indicators spanning health, education, and standard of living (Box 9.1). Drinking water access is one of 10 indicators making up a composite

Table 9.1 Global numbers of people lacking access to improved and safely managed domestic water services, 2017 data

	Global total (m)	Rural dwellers (m)	Rural % of global
Number lacking even a limited service[1]	604	510	84%
Number lacking even a basic service[2]	831	646	78%
Number lacking safely managed service[3]	2,189	1,597	73%

Note: Global total population in 2017 was 7.55 billion; rural population was 3.40 billion
[1] Includes those using unimproved groundwater sources and surface water
[2] Includes those using unimproved groundwater, surface water, and limited services
[3] Includes all those using services inferior to safely managed
Source: JMP, 2019a

> **Box 9.1 What it means to be multi-dimensionally poor**
>
> The multi-dimensional poverty indicator gives equal weighting to deprivations in regard to health, education, and standard of living. A person is defined as multi-dimensionally poor if they are deprived in at least a third of the weighted indicators (i.e. their total score is at least 0.33)
>
> In regard to health, a person is deprived if they live in a household where:
>
> - an adult under 70 years of age or a child is undernourished (score 0.167)
> - any child under 18 years has died in the previous five years (score 0.167)
>
> In regard to education, a person is deprived if:
>
> - no household member aged 10 or over has completed six years of schooling (score 0.167)
> - any school-aged child is not attending school up to the age at which (s)he would complete class 8 (score 0.167)
>
> In regard to standard of living, a person is deprived if:
>
> - the household cooks with dung, wood, charcoal, or coal (score 0.056)
> - the household's sanitation facility is not improved, or it is improved but shared (score 0.056)
> - the household does not have access to improved drinking water, or safe drinking water is at least a 30-minute round-trip walk (score 0.056)
> - the household has no electricity (score 0.056)
> - housing materials for at least one of roof, walls, and floor are inadequate: the floor being of natural materials and/or the roof and/or walls are of natural or rudimentary materials (score 0.056)
> - the household does not own more than one of these assets: radio, TV, telephone, computer, animal cart, bicycle, motorbike, or refrigerator, and does not own a car or truck (score 0.056).
>
> *Source*: Alkire et al., 2020

score; the chosen cut-off (deprived/non-deprived) is the limited/basic distinction in the JMP and Sustainable Development Goal (SDG) definitions. In other words, those with a round-trip travel time of more than 30 minutes are defined as deprived.

Cross-referencing the MDPI database (OPHI, 2020) with the World Bank's (2020c) list of low-income countries suggests that about 350 million rural people and another 55 million urban-dwellers in low-income countries experience multi-dimensional poverty (Table 9.2). Nearly three-quarters of the rural population of low-income countries suffer multi-dimensional

Table 9.2 Numbers of people in low-income countries experiencing multi-dimensional poverty

Setting	Total population (m)	Population in multi-dimensional poverty (m)	Poor as % of total
Urban	239	55	23%
Rural	476	350	74%
Totals	715	404	57%

Source: OPHI, 2020

Table 9.3 Numbers of people in lower-middle income countries experiencing multi-dimensional poverty

Setting	Total population (m)	Population in multi-dimensional poverty (m)	Poor as % of total
Urban	1072	112	10%
Rural	1859	694	37%
Totals	2931	806	28%

Source: OPHI, 2020

deprivation or poverty. The situation in lower-middle income countries is more muted, but the rural-urban imbalance remains (Table 9.3). In short, more than 1 billion rural people in low-income and lower-middle income countries experience multi-dimensional poverty.

Disadvantage within rural communities

We live in a deeply divided world, where significant numbers of people in all countries experience exclusion (from services and decision-making processes) and unequal opportunities. Women and children, people of certain skin colour or racial/ethnic origin, people with physical or mental disabilities (to name only a few) all routinely experience such discrimination. A vicious circle of ingrained beliefs and systemic factors (including laws and constitutional clauses, or their absence; law enforcement and its absence) perpetuates their marginalization. In order to change the status quo, changes need to take place in access to assets and the services they can provide (the subject of this book); but it is also necessary for the voices of those who experience discrimination to be articulated; and for public policies and attitudes to change, too. When all three of these areas are addressed, based on a sound understanding of who is being left behind and why, a more equitable society could be possible.

Many groups in society are marginalized, excluded, or by-passed in development processes. Some are socially excluded (including women); some are poor or economically excluded; others are particularly vulnerable (for example because of location or age) (IDPG, 2017). They include:

- those who are remote from the capital or the district headquarters, and consequently hard to reach;
- those pastoralists and others whose non-sedentary lifestyle renders them hard to serve;
- groups in society that are particularly poor;
- ethnic, caste-based, religious, and other minorities (for example LGBT+ people) who lack influence in society or who are discriminated against;
- women and girls, whose status is often considered to be inferior;
- people with mental or physical disabilities;
- people with chronic illnesses (e.g. HIV/AIDS);
- the various age groups – children, youths, and older people in particular.

Within communities, the issues that reinforce gender disparities and discrimination against those with different abilities or people who are seen as 'other' are not necessarily very different in rural areas than in society more widely. In rural areas women's disqualification from owning title to farmland may be particularly detrimental. Furthermore, they usually have heavy responsibilities for unpaid care work and physically burdensome and harmful tasks such as hauling of water and firewood. Discriminatory attitudes to people with HIV/AIDS, mental illnesses, and other diseases and impairments, can be strong.

The pace of change – breaking down inequalities – may be higher in urban areas than in more traditional rural communities. On the basis of research in Zambia and Cambodia, Evans (2018, 2019) suggests that city life may disrupt gender inequalities because (a) women discover new interests, (b) they see other women demonstrating equal competence to men, and (c) they meet, discuss, debate, and learn together. If this is true more generally, then one of the beneficial impacts of greater interchange between rural and urban areas as urbanization proceeds is that discrimination and inequality may reduce. However, as Boudet et al. (2013) discovered, traditional gender roles that associate men with income generation and household authority, and women with homemaking and childrearing, remain pervasive across all 20 countries in which they conducted their study, with only some 'softening' of norms in urban areas. Consequently, a reduction in pervasive inequalities due to increased urbanization should not be taken for granted.

Rural attitudes and practices

Since the 1980s, it has been evident that rural communities were willing to make very significant efforts towards the realization of their own water services. Although most of the contributions so provided have been in-kind, in the form of labour and the provision of locally available materials, this has been key to progress to date.

Furthermore, although development practitioners and academics alike routinely bemoan low levels of rural waterpoint functionality, condemning community-based management as ineffective and obsolete (see Chapter 7), perhaps such judgments are at least to some extent misplaced. Given the skills and resource limitations in low-income rural contexts, and the near-absence of management support, it is arguably unlikely that alternative management models would have worked better to date. Rural communities have done what they can, and the results, though disappointing, are not catastrophic.

However, some practices of rural water users, and the underlying attitudes that inform them, undoubtedly weaken efforts to both provide sustainable domestic water supply services and to ensure the full enjoyment of the benefits such services can bring. For example, the unwillingness of some people (perhaps a minority) to pay for reliable water services, and slow progress in regard to adopting good hygiene behaviours, exemplify this point.

Demographic trends in rural areas

As the wealth of nations grows, and as urbanization continues, a demographic transition takes place in which fertility and mortality rates fall, the rate of population growth slows, and the population ages (Bongaarts, 2009). In the extreme, growth turns negative and population size decreases. While the wealthier regions of the world have already gone through the demographic transition, Africa, Asia, and Latin America saw the process start later, and it is still continuing.

Figure 9.1 shows the projections of rural populations in low-income and lower-middle income countries to the mid-21st century. Although the population of the lower-middle income countries is expected to peak in the 2030s and then gradually fall, the continuing rise in the low-income countries (more than three-quarters of which are in sub-Saharan Africa) is clear.

In those regions and countries where urbanization is happening rapidly, this adds further complexity to the rate and nature of rural population change. While rural populations continue to reproduce, many of the younger, more mobile, and arguably more ambitious members of rural communities make the journey to towns and cities in search of work and other opportunities.

Much of the literature on urbanization understandably focuses on its impacts on cities themselves and on people living in unplanned settlements or 'slums'; far less discusses the corresponding implications for rural areas. However, it seems likely that as predominantly youthful migrants leave rural areas, the communities left behind may contain greater proportions

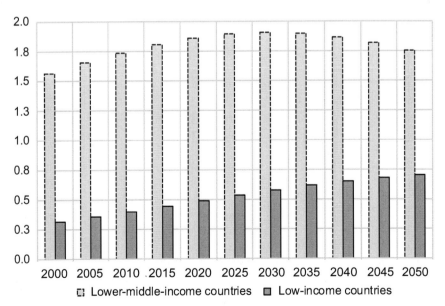

Figure 9.1 Rural population growth and projections 2000–50, billions
Source: UN DESA, 2019b

of women, older people, and people whose disabilities prevent them from migrating easily (Anríquez and Stloukal, 2008). This has implications for the management and financing of rural water services, although, on the upside, increased remittances from family members working in towns and cities may put more money in the hands of rural households (McKay and Deshingkar, 2014) – some of which could help to pay for maintenance and repairs of rural water infrastructure.

Assessing the effectiveness of rural water supply programmes

Rural communities in low- and middle-income countries generally suffer high levels of multi-dimensional poverty (inadequacies in health, education, and standard of living), which pose numerous challenges to their survival and ability to flourish. And yet they also exhibit effective and sophisticated self-government, organization, and capabilities, all of which are too frequently overlooked by external programmes.

The following paragraphs are more of a personal reflection on four decades of work in the rural water sector than a literature-based analysis of the performance of the sector. I draw attention to three aspects in particular of rural water programmes. Each has serious implications for rural communities' experiences of water services. All can be addressed, but only with a step-change in personal, professional, and political commitment.

First, in their engagement with rural communities, water programmes tend to apply a one-size-fits-all approach. The formula involves some form of sensitization of the community (informing and enlisting communities in subsequent activities in which their role will be important), the establishment of a water user committee (also known as water point committees), some training for that committee, and the handover of the water infrastructure for management and subsequent financing by the committee. In the case of piped systems, the community management organization is correspondingly more complicated, but a formula or set of steps still tends to be applied. The procedures for engaging with communities and other stakeholders are often set out in a formulaic manner (see, for example, the 'software steps' by the Ministry of Water and Environment (Uganda), 2012). Community engagement, participation, and management are reduced to standard steps, and this inevitably deters programme implementers from taking full account of the particular features – the size, cohesion, leadership, location and connectedness – of each community. As already noted (Chapter 4), rigid adherence to standard designs of physical infrastructure may often undermine its subsequent performance. Similar considerations apply in the case of the social dimensions of water programmes.

The second flaw in many rural water programmes is arguably more serious still. This is the failure to actually observe known good practices. Much experience in the sector has led to an extensive body of learning about what works and how to be effective. Taking shortcuts in the social aspects of rural water programming is no different in principle to the corresponding cutting of corners in engineering. Investing insufficient time and individualized attention to the social aspects of rural water programming is as likely to undermine the subsequent performance of the service as, for example, using insufficient cement in a concrete mix, or using galvanized steel pipes in corrosive groundwaters. In a study of water service sustainability in southwest Uganda (Carter and Rwamwanja, 2006), one of the main findings about the programme's high levels of effectiveness and sustainability was its refusal to merely pay lip service to sound principles of community participation and management (Box 7.4). This remains true to the present day.

Third, even if the community is effectively engaged at the outset of a programme, it is rare for there to be any significant follow-up and monitoring of community management and its effectiveness. Little or no post-handover support is provided by the local government or NGO that implemented the programme.

The negative impacts of poor quality programming

Rural water programmes that fail to ensure the requirements for continued functioning of the water supply represent not only wasted investments, but at least as importantly, serious negative impacts on rural people. The UPGro 'Hidden Crisis' project examined such impacts in Ethiopia, Malawi and Uganda, and I am indebted to Luke Whaley, social science researcher on that project for the following observations.

Poorly functioning and non-functional waterpoints affect rural people in a range of ways. Most directly, a non-functioning waterpoint often results in community members having to use less safe or unsafe sources (e.g. contaminated hand-dug wells, swamps), travel further for water, pay for water (e.g. at a kiosk), or place additional demand (and strain) on alternative sources. Those who are healthy and physically fit, have the financial means, and/or have a means of transport (e.g. bicycle or motorcycle) can more easily access safe alternative water sources when a waterpoint breaks down. Those who are elderly, have a physical disability, lack the financial means, or have no access to transport may suffer considerably. This is especially the case if these people do not have relationships with others (lack the social capital) who may provide for them during this time.

The UPGro research revealed that it was common for neighbouring villages to have a reciprocal relationship whereby the inhabitants of one community would use the waterpoint of another when their waterpoint became non-functional (a regular occurrence). This may strengthen the bond between villages. However, depending on the time of year (e.g. during the dry season) and the existing demand on the functioning waterpoint, this can cause social tension and conflict even if the principle of reciprocal access is accepted by many or most.

Waterpoint breakdowns can actually build the capacity of community water management arrangements, especially if they are relatively minor.

This is because the management arrangement develops the structures, processes, and sources of authority needed to undertake repairs. On the other hand, a waterpoint that repeatedly breaks down despite the best efforts of the community, or that performs poorly due to fundamental issues with, for example, siting, depth of borehole, or aquifer properties, can undermine management capacity. This can lead to abandonment of the waterpoint.

Poorly functioning waterpoints – especially relating to quantity – can result in long queues and a long waiting time, typically for women and girls. Long queues also occur at well-performing waterpoints due to high demand, which is exacerbated when other sources in the locale become non-functional. This loss of time has obvious knock-on effects relating to domestic work, education, employment, and leisure time. Long queues also tend to result in increased tension and conflict at the waterpoint. This may be both verbal and physical, and can cause wider rifts between individuals and households in a community.

Moreover, there is a strong gender dimension to the time spent fetching water. Often husbands become suspicious of wives who are away for long periods of time (suspecting them of adultery). The ramifications of this are again tension and conflict. We even have evidence of physical beatings inflicted by husbands on wives who have had to queue for hours on end at a waterpoint that is poorly performing or in high demand, or who have had to travel long distances to collect water.

In the context of the community management model for water, another sticky issue is finances. The need to raise funds from water users for repair work leads to much suspicion and conflict, and fallouts are common. This centres on who raises funds and how the money is spent.

All of these observations make it imperative that organizations implementing rural water programmes do so in ways that ensure continued functioning and sustainability. Part of this is about the way the community is engaged in the first place, but the whole picture is much wider, requiring on-going support from the mandated agency external to the community.

Doing rural community water supply well

The final section of this chapter addresses two of the key issues that all rural water programmes need to get right: first, how to 'do' community participation well; and second, how to ensure, as far as possible, that the benefits of rural water programming reach everyone, regardless of where they are or who they are.

Community participation

The literature on community participation in rural water supply has been clear over many decades about the necessity and potential benefits of free, meaningful, and inclusive participation of communities in their own development. As pointed out by Jiménez et al. (2019), target 6b of the SDGs – 'support and strengthen the participation of local communities in improving water and sanitation management' – is the only target to mention public participation as a key means of implementation.

Harvey and Reed (2007: 367–9) usefully distinguish between community participation (in planning and decision-making) and subsequent community management of water services, arguing that the former 'is a pre-requisite for sustainability, i.e. to achieve efficiency, effectiveness, equity and replicability', while the latter is not. Like many in the management debates around community water services, the authors are critical of the community management model, and are of the view that its time is over. The authors identify six reasons why communities often struggle to manage their water services well. These include the reliance on voluntary inputs; the absence of procedures for replacing committee members; issues of trust; failure of water users to pay agreed tariffs; limited or no contact with local government; and communities' inability to replace major capital items (Chapter 7).

When it comes to the practice of community participation, most of the literature consists of manuals, toolkits, and how-to guides, which, although written with the best of intentions, can discourage thoughtful analysis by practitioners of the particular features, strengths, weaknesses, and ways of working with individual communities. In contrast, Wisner and Adams (2002) provide a list of six important principles for community engagement. These are set out briefly in Box 9.2.

Kleemeier (2000), in her study of the Malawi piped water schemes, tabulated the conventional features of community participation activities in rural water development, explaining their rationale. She viewed them as hypotheses rather than demonstrably key features of successful programmes. The list of features is reproduced here as Box 9.3; the rationale is set out in Table 1 of Kleemeier's paper (2000: 932).

Reaching everyone

The summary idea here is equality – not equality of treatment, but equality of outcome. In other words, that irrespective of individual or group characteristics,

---

Box 9.2 Some key principles regarding community engagement

Find an entry point to the community, possibly an existing community-based organization or the local primary health care system.

Work with community leaders, those, including women leaders, who are able to bring people together and promote action.

Ensure official support for community-led projects.

Understand the socio-economic make-up of the community, including its divisions, and its past history of self-help community projects (especially if these have failed).

Make special arrangements to encourage participation.

Understand the existing community formal and informal organizations.

Source: Adapter from Wisner and Adams, 2002, Chapter 15

> **Box 9.3 Typical features of rural community water projects**
>
> - Meetings to explain project before it begins, community has right to refuse it.
> - Contract signed specifying community's and project's responsibilities.
> - User committee formed with design and construction responsibilities.
> - Incorporate local preferences and knowledge in choice of technology, design, and construction.
> - Same committee or new one assumes operations and maintenance (O&M) responsibilities.
> - Community upfront cash collection (to contribute to capital costs, establish O&M fund, or both).
> - Community provides free labour and other materials.
> - Management and book-keeping training provided to committee members; management procedures established.
> - Technical training and tools provided to local repair persons.
> - Hand-over ceremony.
> - A staff of community mobilizers to carry out the above activities.
> - Simple technologies (VLOM handpumps, gravity schemes, protected springs, etc).
>
> *Source*: Kleemeier, 2000

all people experience the benefits of sufficient, safe, and sustained water services. Governments, donors, international agencies, and NGOs are rightly concerned that the benefits of rural water services should reach everyone, irrespective of gender, age, disability, health, ethnicity, religion, or social status.

What do equity, equality, and inclusion mean?

In WaterAid's 'Equity and inclusion framework' (Gosling, 2010) a distinction was made between equity – 'the principle of fairness' – and inclusion – 'the process for ensuring that all are able to participate fully'. The pursuit of equity requires a recognition of the different needs of different people, and actions to compensate for discrimination and disadvantage. The process of inclusion involves 'supporting people ... to engage in wider processes to ensure that their rights and needs are recognised' (Gosling, 2010: 6). According to Gosling, achieving equity and inclusion requires:

- better recognition and understanding of the differential needs of individuals and groups;
- identifying and tackling the root causes of exclusion;
- promoting and supporting their inclusion in decision-making processes; and
- identifying and implementing appropriate and sustainable solutions.

For Water Aid, the pursuit of equity and inclusion is inextricably linked to a rights-based approach to development. The human right to water and sanitation was recognized in 2010 by the United Nations General Assembly and the Human Rights Council. In 2013 the Human Rights Council agreed on the content of this right (which refers to availability, quality, accessibility, affordability, and acceptability of services), and many nation states have incorporated the human right in their constitutions and national legislation.

Much of the subsequent literature on the rights to water and sanitation by the UN Special Rapporteur and others has devoted significant space to explaining what the rights to these services are not (for example, to be enjoyed immediately or free of charge). The human right is based on considerations of non-discrimination and equality, which have to be explained in a similar way. For some organizations the term 'equity' has been replaced by 'equality', which, according to de Albuquerque, is explained as follows:

> Non-discrimination and equality are linked under human rights law: States must ensure that individuals and groups do not suffer from discrimination and that they can enjoy full equality. Equal does not mean "the same" nor "identical treatment in every instance". Human rights law requires equal access to basic services, but this does not mean that everyone must benefit from the same technical solutions or the same type of service, such as flush toilets. Equality does not imply treating what is unequal equally. People who are not equal may require different treatment in order to achieve substantive equality. States may need to adopt affirmative measures, giving preference to certain groups and individuals in order to redress past discrimination. (Albuquerque, 2014, Book 7: 12)

In the desire to see water, sanitation, and hygiene (WASH) services reach all, it is of course not necessary to embrace the human rights perspective explicitly, and many governments and NGOs only make passing reference to it, if at all, while still endeavouring to bring sustainable services to all. In short, the goal of WASH services for all, regardless of race, colour, sex, language, religion, political or other opinion, national or social origin, property, birth, or other status (to quote the Universal Declaration on Human Rights) is one which all can share. The questions addressed here focus on how to achieve it in practice.

Table 9.4 Achieving equality in rural water programmes

Planning and design stage

Selection of target population

There can be an understandable tendency to select target populations whose needs are easy to meet. Rather, at least part of programme resources should address the needs of hard-to-reach groups.

Understanding the target population

The nature of the groups within the target population that have different needs must be identified through some form of baseline study or formative research.

Participative design for inclusive access

Selection and detailed design of technologies for rural water supply must be undertaken with the full participation of those that these services are meant to benefit. Accessibility and safety audits (WEDC/WaterAid, n.d.) can be highly effective here.

Implementation

Taking participation seriously

All groups within the target population must be facilitated to actively participate in decision-making processes, to the full extent of their willingness and ability to be so involved. Given the particular roles of women in rural water supply, their representation must be especially full and meaningful. Participation must be fully inclusive.

Addressing specific needs of identified groups

There are many barriers to access by disabled people, older people, those with HIV/AIDS and other chronic illnesses, and minorities. Environmental or physical barriers to access must be addressed through accessible designs and meaningful participation of affected groups.

Working to change discriminatory attitudes

For equality to become a reality, programmes must work to alter discriminatory attitudes. Whether these are (male) attitudes to women, prejudices about mental illness or HIV/AIDS, racist or other discriminatory attitudes, programmes must identify and work with others to challenge and change them.

Addressing institutional barriers

Working with others to strengthen laws and policies

The national legal framework may fail to protect minorities from discrimination, and water and other sector policies may neglect to explicitly include all groups who need water services. While it cannot be the work of one organization or programme to change these, there is an onus on all water programmes to collaborate in the strengthening of the legal and policy framework. Organizations representing people who tend to be excluded should be involved in these processes.

Contributing to and disseminating guidance

Guidance material emanating from national policies needs to accurately reflect equality and inclusion concerns, and it needs to be put in the hands of those whose day-to-day work applies it. Programmes must play their part in the formulation and dissemination of such guidance.

Developing capacity in key institutions

Professional knowledge and attitudes, and the culture of organizations, must reflect equality and inclusion concerns applied among target populations. Where discriminatory practices exist (e.g. toward women or disabled people) programmes should collaborate and challenge such attitudes.

Monitoring

Keeping track of access by identified groups

Monitoring of access to and use of rural water services should be disaggregated by the main groupings identified in baseline or formative research.

Monitoring social attitudes

Some measure should be attempted of trends in social attitudes toward marginalized groups, especially if programmes undertake specific actions to address such attitudes.

Monitoring progress in overcoming institutional barriers

Monitoring of attitudes and practices within organizations, as well as the wider institutional framework of laws, policies, and guidance (and their implementation) must be attempted.

What types of action deliver equality and inclusion in WASH?

Addressing equality and inclusion requires action at multiple levels. Projects and programmes have target populations, all of whose WASH needs must be identified and addressed. Discriminatory attitudes towards those who are commonly excluded also need to be addressed, widening the circle of actions beyond the target population. Government policies, guidelines, and budget allocations also need to reflect a positive attitude to meeting diverse needs.

In this vein Gosling (2010) highlights three barriers to inclusion that must be overcome:

- **negative attitudes** such as prejudice, pity, isolation, overprotection, stigmatization, misinformation, and family shame;
- **environmental barriers** including difficulties of physical accessibility and those aspects that present particular difficulties for specific groups (e.g. access to visual communications by those with visual impairments);
- **institutional/organizational barriers**, including policies and processes that exclude or neglect those with particular needs, whether deliberately or not.

All three of these barriers have to be addressed in order to achieve truly equal and inclusive outcomes. The first and third are arguably the most difficult and time-consuming to change since they involve cultural and institutional change. The second can at least be addressed in the short-term through programmes that aim to reach the entirety of their target populations. For example, recent guidance on WASH technologies that are accessible to disabled and elderly people has been published (Jones and Wilbur, 2014).

What should projects and programmes do to assure equitable and inclusive outcomes?

Building on the three broad areas of action outlined above, Table 9.4 sets out some key actions that should be included in WASH projects and programmes to bring about equitable and inclusive outcomes. Taken together, these actions help to address the negative experiences of marginalized people, whether the causes of those experiences originate from societal attitudes, in the ways that rural water programmes are typically implemented, or from policies and institutions that result in systematic disadvantage.

CHAPTER 10
Water for all: why is it such a struggle, and what can be done?

Abstract: *Getting a decent supply of domestic water to everyone has proved to be much more difficult than the architects of the first United Nations drinking water decade (1981–90) ever imagined. This chapter explores why, and outlines some of the ways in which obstacles to progress have been addressed in recent years. The main focus is on the people involved in the communities, governments, and non-government organizations who work to extend water access. They are also central to the problems faced by the water sector, and hence fundamental to change and progress. Within rural communities, limited willingness to pay for services, negative attitudes to minorities or 'others', and inadequate hygiene behaviours all limit the achievement of rural water sector goals. In the organizations working to extend and improve water access, corruption, poor quality work, lack of leadership, and limited accountability all stifle progress. Poverty within communities, and scarce resources in the sector also put the brakes on effective development. Finding ways of enabling and encouraging constructive individual behaviours while simultaneously hindering negative behaviours can contribute to positive change. These include: generating community level demand through citizens' action and through education around the human right to water; working to improve governance and democracy; undertaking participatory sector analyses and financial tracking exercises; carrying out sustainability checks; promoting professionalization of sector actors; reforming institutions; and fighting corruption.*

Keywords: accountability, citizens' action, human right to water, bottleneck analysis, sustainability checks, financial tracking, professionalization, institutional reform, anti-corruption, systems strengthening

> 'Pay attention to what is important, not just what is quantifiable'
> —Meadows, 2008

Introduction

Our planet generally has more than enough freshwater to supply its entire population. If nations allocated sufficient financial resources and invested in the needed education and training, it would be possible to engineer water supply systems that would provide enough water of adequate quality in a reliable and affordable manner. We know a great deal about how to manage and finance rural water supply. And yet, despite decades of investment and effort, safe water for all remains an elusive goal. The Sustainable Development Goal (SDG) target 6.1 – 'by 2030, achieve universal and equitable access to safe and affordable drinking water for all' – will not be met without fundamental changes to how the water sector works, and how quickly it can respond to the global needs. At the time of writing this book, there is less than one decade remaining to achieve the SDG targets, and the world is seriously off-track in realizing its ambitions with regards to safe and sustainable drinking water for all.

In this chapter, I focus first on why I believe the world is collectively failing to serve some of its most deprived people with this basic necessity. I then indicate in general where the solutions may lie. My framing of the problem in this chapter is more reflective of my personal journey and views than some of the material in the other chapters. However, wherever I can, I provide evidence to support my assertions and opinions. The final two chapters of the book go into greater detail and specifics as to the ways forward; this chapter may therefore be considered as a preface to the challenging topic of creating systemic and lasting change in rural water supply.

Root causes, generic solutions

There are many answers to the question of why greater progress has not been made in sustainable rural water provision: insufficient prioritization of the (sub-)sector, inadequate allocation of budgetary resources, ineffective leadership, lack of regulation and enforcement of policies, poor management of resources, lack of coordination of disparate actors – the list goes on. Generic solutions to the limited progress have predominantly focused on addressing different fundamentals in broad ways, some of which are described later in this chapter. At the time of writing, the approach in the sector that is gaining most traction is that of 'systems strengthening' (Huston and Moriarty, 2018a; WaterAid, 2019), the concept being that multiple elements of a 'system' comprising numerous building blocks need to be addressed to bring about the required systemic change. For example, WaterAid contends that:

> System strengthening means understanding that WASH [water, sanitation, and hygiene] exists in complex systems with many component parts and within different social, economic, political and environmental contexts. It involves identifying and working to address the barriers in behaviours, policies, processes, resources, interactions and institutions that block achievement of inclusive, lasting, universal access to WASH. (WaterAid, 2019: 1)

I will return to potential solutions, including that of systems strengthening, but first, without understanding the roots of the problem, it is unlikely that sound solutions can emerge and be intelligently assessed. The following paragraphs paint a picture of the problem, in order to subsequently explore better ways of working.

Systems and individuals

Huston and Moriarty describe the system in these terms: 'We use WASH system to describe all the people, components and functions that are needed to deliver WASH services. The WASH system includes all the **actors** (people and institutions) and all the **factors** (infrastructure, finances, policies and environmental conditions) that affect and drive the system' (2018a: 7, emphasis added). My starting point here is that not all actors and factors are equal in terms of how they influence the working of the entire system. If we imagine the system in low- and middle-income countries to be like a rather old and at times unreliable car, then arguably its driver is key – a driver who knows the idiosyncrasies of their vehicle may be able to coax it over difficult roads. A driver who is unfamiliar with the vehicle's faults may be less successful.

Just as with that car, at the root of the rural water system are its drivers – individuals who possess or lack knowledge and wisdom; people who hold ideas, beliefs, biases, and prejudices; and who act, or fail to act, in a variety of ways that result in either progress, stagnation, or worsening of the water services enjoyed by all citizens. The roots of the **problems** faced by the water sector, and equally the possibilities of **solutions** to those problems, lie squarely with the people – the individual men, women, and children – who create and shape institutions, who design laws and policies, who allocate funds, who build facilities and train water users, and who consume water. Whether we are considering water users, staff of implementing agencies, those employed by funding organizations, decision-makers, policy-makers, researchers, consultants, or those in campaigning groups, many negative individual attributes are apparent – all of us are flawed in many ways. Our individual attributes help to shape, and at the same time are shaped by, social norms as well as organizational and societal cultures.

The mixed motives, ambiguous attitudes, abuses of power, and inconsistent effectiveness of flawed human beings are an important reminder that the systems people create – governments, economies, legislatures, civil services, civil societies, private sectors – also contain in-built flaws and strengths. Individual values, and the collective ethos that individuals build and reinforce, really matter. People are at the root of the problem of erratic progress toward safe and sustainable water for all; but people are also the solution. If we can design into the system ways of *hindering* those aspects that constrain progress, and *enabling* those that make possible consistent and steady progress in the right direction, then even otherwise imperfect systems may become more effective.

For example, in their paper on the achievements of the Kigezi Diocese Water and Sanitation Programme (KDWSP) in Uganda, Carter and Rwamwanja (2006) refer to the importance of the individual and organizational ethos of those working for the betterment of society. They highlight how a strong passion for the work, driven by compassion for the marginalized, driven by humanitarian or religious motives, or motivated by strong instincts for social justice and equity, can make the difference between success and failure.

The individual and organizational values of KDWSP result in a commitment to undertaking high quality work, and a refusal to cut corners or engage in corrupt practices. Such situations are not widespread, but where they are found they should be celebrated.

Because important aspects of work in the water sector involve attempts to change attitudes, values, and behaviours of people and wider society – water users, the organizations that serve them, and those who make key decisions – such change must begin with you and me. This has been put well by Cavill et al. (2020) in a call to action around individual, professional, and organizational change for gender-transformative WASH programming. I stress this as a starting point, while recognizing that organizational and wider contextual factors beyond the individual's control commonly limit what is possible.

Causal explanations

Explanations as to why a certain state of affairs exists can be made at different levels. A handpump ceases working because a rubber cup seal tears and disintegrates. Why did that happen? Perhaps because it was of poor manufacturing quality and it failed prematurely. Why was a poor quality spare part used? Maybe because the country has no system for routine inspection and quality assurance of handpump parts. Why is this so? Perhaps because of limited government budgets, as a consequence of failure to recognize the importance of such quality assurance work. And so on.

In some cases, asking a succession of 'why?' questions can help to move explanations from the immediate causes to some of the underlying systemic reasons why things happen the way they do. In other cases, much more tangled webs of inter-related causes add to the complexity of causal analysis. The common thread, however, is that the deepest reasons as to why the rural water sector lacks effectiveness are often attributable to the national and local context, and decisions that are taken at government level regarding budget allocations and ways of working.

The wider system

I have started by focusing on individuals and their qualities. Individuals build households, communities, institutions, and, by extension, society. These collective expressions of cultures and nations compose what is more commonly meant by the wider system. This wider context is sometimes referred to as 'the enabling environment'. When water sector actors work to change or influence the enabling environment, they are engaging in what has become variously referred to as systems strengthening (as noted above), systems support, or, in UNICEF's case, 'upstream work' (in contrast to its direct 'downstream' work to improve water access in communities) (UNICEF, 2018).

The water sector in a nation, and the various actors who work in it, are moulded and constrained by the higher-level contextual issues of history, geography, politics, and economics (among others), and in turn these actors and factors (to use Huston and Moriarty's terms) affect the services enjoyed by communities. However, the level and quality of those services are also affected by water users themselves – their motivations, commitment, and willingness to take on responsibilities for their own services.

Some important components of the system

Behaviours at community level. It would be naïve to think of water users as always compliant, responsible, and practising behaviours that reinforce external attempts to improve water services for all. Inability and/or unwillingness to take on responsibility to pay for improved water, or to play a part in the management of water supply, or indeed both, can constrain what is possible. Attempts to roll out private operator or rural utility models of water supply show the difficulty of persuading some rural households and communities to pay even for demonstrably high-quality services. Reluctance of individual households and communities to sign up to such arrangements may very rationally be based on genuinely limited disposable incomes, lack of trust in private entities and public sector utilities, or poor past experiences of the promises made in the name of 'development'.

In all rural communities, some households are truly unable to pay even very modest water charges. In many cases they are known to the community, and they are exempted from payment. In some cases, they may receive less sympathetic treatment, which may in part be motivated by prevailing discriminatory attitudes on the basis of gender, ethnicity, or other socio-economic status. Even where some households are exempted from payment, rarely do wealthier or more able community members take on the additional financial responsibility that, logically, should accompany otherwise reduced overall revenues. In England and Wales, for example, where the law prevents a household's water being cut off by its provider for non-payment, it was estimated that in 2014–15 the total amount unpaid was about £2.2 billion. This had the effect of adding £21 per year to the bills of paying customers – about 5 per cent of an average household's bill at the time (Priestley and Rutherford, 2016).

In terms of water use behaviours, the potential public health benefits of rural water supply are probably determined more by hygiene practices such as hand-washing with soap (as well as by food hygiene) than by the quality of water that is consumed as drinking water. The evidence, limited though it is, suggests that hand-washing with soap is very far from being universally practised. According to the Joint Monitoring Programme (JMP, 2019) around 40 per cent of people globally in 2017 had only limited hand-washing facilities (lacking water or soap) or no facility at all. Many WASH professionals would argue that a fundamental part of the problem is the limited 'voice' and demand from people whose water services are inadequate. Citizen and community scorecards (Post et al., 2014; Ryan, 2008) and the promotion of the human right to water (UN, n.d., b) are attempts to address this aspect of the problem.

Governments and implementing agencies. Together, national and local governments and their funding partners, non-governmental organizations (NGOs), faith-based organizations, academics, consultants, and private entities work to improve and sustain water access. Each has its strengths and weaknesses. We will come to the strengths shortly, but on the negative side (and at the risk of presenting something of a caricature), the following constitute some of the main criticisms:

- Governments can seem over-bureaucratic and they are often under-resourced.
- Some NGOs and faith-based organizations operate in unaccountable ways, failing to fully comply with national policies.
- When rural water supply is seen as a charitable endeavour by NGOs rather than the duty of the state, this can let governments off the hook and lead to those governments failing to commit the needed financial resources to the work of sustaining public services.
- Some academics may be overly focused on publication impact and personal prestige.
- Some consultants may be insufficiently rigorous in their generation of evidence and advice.
- The private sector is often accused of being too pre-occupied with profit.

All of these tendencies (or stereotypes), and others, arise in part from the attributes of individuals, and in part from the cultures of the organizations involved – the norms that are created through the aims, procedures, and impact metrics with which they have to comply. Individuals in the system are interested, to a greater or lesser extent, in monetary reward, personal comfort and security, and recognition. None of these things need have unduly negative consequences, unless and until they exceed certain boundaries.

Corruption. One of those boundaries, and one which extends well beyond the water sector, is corruption. The individual and systemic dimensions of corruption are well known, in their negative impacts on people's ability to access public services and on the quality of those services. Once corruption becomes endemic and normalized, it is extremely difficult for individuals to rise above it. As Brioschi (2017) points out, when the fact of corruption crosses the line from open acknowledgment to public denial, then a systemic problem has become further entrenched.

Brioschi takes a very pragmatic and matter-of-fact approach to the subject of corruption, cataloguing its history and geographical extent globally – and demonstrating its pervasive presence. It is neither new, nor particularly concentrated in low-income countries. The excesses of African dictators in the

20th and 21st centuries are easily matched by European and North American abuses of power and privilege, and concentration of wealth. And yet even Brioschi admits that 'corruption ... demands a de facto tax on corporations and individuals that diverts resources from the public good' (2017: 207). Something has to be done. Organizations such as Transparency International and the Water Integrity Network are at the forefront of attempts to expose and oppose corruption generally, and in the water sector specifically. Their work is mentioned again below.

Quality of work. In the absence of effective supervision, oversight, and accountability, it is all too common for the quality of construction work to be mediocre, shoddy, or completely unfit for purpose. Implementing organizations may be doing the right things, but if they are not done well, the results cannot be expected to last. Poor quality of construction, and inadequate attention to thorough processes of community engagement, stand out as especially important in rural water supply.

The incentives to cut corners are clear – for the individuals involved, it is often easier or more comfortable, though far less professionally satisfying – to do poor quality work, and for contractors and consultants it often leads to greater financial gain. Such tendencies have to be actively countered, for example by rigorous supervision practices, by strong engagement of the communities for whom work is done, and by exposing poor practices where they exist.

Leadership and accountability. The systemic ineffectiveness that is all too common in rural water development often results from shortcomings in leadership and an absence of real accountability. Individuals operate within community structures or organizations external to the community, and it is up to those leading these bodies to ensure their effectiveness. As Wenar (2006: 5) has stated, 'Responsibility must be fulfilled, and responsibility must be seen to be fulfilled'. Ultimately it is those who lead – in governments, donors, NGOs, private companies, and other actors – with whom the accountability buck stops.

Making leadership more accountable and more effective involves a balance of education and persuasion, as well as processes (for instance involving citizen scorecards and other feedback mechanisms) that require action and improved performance (see below).

Towards solutions

It is important to seek ways of countering the negative beliefs, attitudes, biases, and prejudices of individuals; the behaviours that these underlying attributes give rise to; and their expression in and through institutions (the formal and informal rules of the game) and organizations (the players or actors). At the same time, it is necessary to find ways to encourage, reinforce, and

158 RURAL COMMUNITY WATER SUPPLY

institutionalize constructive and positive tendencies. The broad approaches described in the following sections are intended to bring about changes to the status quo, through actions that put pressure on those organizations that directly provide water services to communities.

In an ideal world community demand would complement corresponding pressure from national governments and their partners. In many low-income countries, however, it is those very same national governments that need to be pressurized, rewarded, or even shamed into fulfilling their mandates. This is where the international community of donors, United Nations agencies, international NGOs, and the media fit in, providing evidence, comparative performance data and analysis (such as the regular JMP reports), and campaigning (resulting in a blizzard of calls to action).

In Figure 10.1, which indicates the various pressures for institutional change, all the arrows (indicating the direction of pressure or influence) should be two-way to show that each entity or group of entities is listening to the others. In particular the link between communities and the international organizations (the large grey arrow) is arguably the weakest, while simultaneously being the most important.

It has become common to call for more 'listening to the poor'; an internet search of the phrase reveals how many organizations consider it important. However, few do more than pay lip service to such an idea. Two important exceptions to this (perhaps slightly unfair) generalization are the World Bank's *Voices of the Poor* studies of the early 2000s (Narayan, Chambers

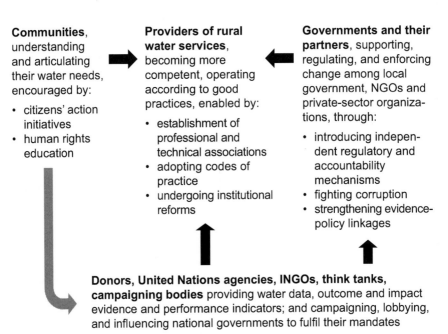

Figure 10.1 Local, national, and international checks and balances

et al., 2000; Narayan, Patel et al., 2000; Narayan and Petesch 2002), and the Collaborative for Development Action study *Time to Listen* (Anderson et al., 2012). These documents demonstrate the huge importance – only matched by the challenges involved – of really hearing what those on the receiving end of development interventions (with or without foreign aid) think of what is done 'for their own good'.

Addressing negative individual attributes

Negative individual tendencies can be opposed by a range of actions, which together place sound evidence, the equal valuing of all people, and a constructive ethos at the centre of all that we do to bring equitable and sustainable water supply to rural communities (Figure 10.2).

Individual attributes and behaviours can be sought out and rewarded with public recognition in various ways. There is much evidence that demonstrates individuals often value recognition from society and from their peers more than monetary reward. As the 2015 *World Development Report* (World Bank, 2015) pointed out, we humans are inherently social and we care about what others think of us (or what we think others think of us). Moreover, we prefer to be 'conditional cooperators' (people who will work with others if we think they are playing by the same rules) than 'free-riders' (looking after ourselves at the expense of our fellow humans). An important part of education, training, capacity development, and professional development programmes should therefore focus on encouraging the motivations, attitudes, and behaviours of the participating individuals – not only their knowledge and skills.

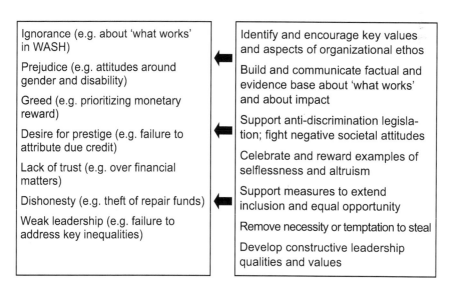

Figure 10.2 Opposing negative individual attributes

Professional and technical associations and institutions exist to promote and enable good practice among their members. They do this through processes of certification by peers, and a requirement for continuing professional development. Naturally such processes are imperfect, but they add another piece to the puzzle of how to raise technical and professional standards. In my view, all technically and professionally educated workers in the rural water sector (and wider afield in development and humanitarian work generally) should belong to at least one such organization appropriate to their field of work. Membership encourages learning and sharing of good practice, and it leads to certification and professional recognition. Why would any responsible individual object to such membership, and why would any potential employer consider applications from individuals who are not so certified (or on the journey to certification)?

From individuals to organizations

Much good work has been carried out in recent years to establish and strengthen national associations of, for example, hydrogeologists, borehole drillers, and engineers. Some of this work has been initiated by external agencies (e.g. RWSN, UNICEF), but there are also good examples of such initiatives being taken nationally, for example in Uganda's water sector.

Research involving human subjects can only be carried out and published if its methods have been scrutinized by ethics committees in the countries where the work originated, and the countries where the research took place.

But much remains to be done, for example in:

- formalizing the certification processes in low- and middle-income countries for water professionals and practitioners;
- codifying practices and procedures, and – equally importantly – communicating these to all players in the sector;
- extending research ethics guidelines to strengthen national leadership and ownership of research programmes – including in publications;
- establishing appropriate and effectively enforced regulatory and accountability frameworks for government and NGOs engaged in rural water provision.

Improving each of these areas requires strong leadership from national institutions, supported by international organizations, which can bring a comparative perspective to the more detailed knowledge held by national water sector technical and professional personnel.

Strategies for improving/strengthening rural water supply

Starting with communities. In the early 2000s, while work was under way to turn the UN's (2002) 'General comment' on the human right to water into the fully fledged rights that were declared in 2010 (for water and sanitation jointly) and 2015 (for water and sanitation separately), a parallel stream of activity was under way to enable the voices of citizens and communities to be heard. 'This is the essence of Citizens' Action: citizens are supported to engage in dialogue with service providers and governments; holding them to account for the provision – or lack of it – of services.' (Swain et al., 2006: 173).

Swain and colleagues drew attention to the discrepancies between 'words and action and between policy and practice ... and between responsibility and action' (2006: 173). The concept behind Citizens' Action was that communities could be supported to articulate their water and sanitation problems, entering into dialogue and engagement with those responsible for providing services, for as long as necessary. These authors and others (Ryan, 2008; Post et al., 2014) described the use of report cards, community scorecards, mapping of services, and holding of public forums and juries to assist communities in analysis and articulation of their situation and demands. Ryan's WaterAid report (2008) outlined 20 Citizens' Action projects in six countries of Africa and Asia. The examples documented show the potential power of strengthening citizens' demands for well-performing services if carried out alongside complementary measures to enhance the performance of those organizations responsible for providing water and sanitation services.

De Asís et al. (2009) describe a number of complementary activities alongside raising citizens' voices in the ways just outlined: having communities actively participate in setting public budgets; making public information on government budgets, expenditures, and performance indicators; and tracking and publishing information on public sector expenditures.

With the declaration of the human right to water and sanitation in 2010, the focus of many organizations in the water sector has shifted from such Citizens' Action programmes, to ways of turning the ambitions expressed in the human rights declaration (Box 10.1) into realities.

The declaration of the human rights to water and sanitation represents globally shared and laudable ambitions (similar to those expressed in the SDGs) for all to enjoy high standards of service as soon as possible. However, human rights thinking and advocacy is not immune to well-founded and constructive critique, especially from legal, economic, and political perspectives – the moral arguments perhaps being less contentious.

This is not the place to undertake a detailed review of human rights thinking. It is sufficient to point out, in the face of the general enthusiasm for human rights-based approaches by UN agencies and international non-governmental organizations, that there is an extensive literature exposing some of the real legal, economic, and political challenges to the realization of economic and social rights, including the right to water (e.g. Neier, 2006; Davis, 2012; Langford, 2017; Ssenyonjo, 2017; D'Souza, 2018; Young, 2019). It is sobering to remind oneself of Article 1 of the Universal Declaration of Human Rights: 'All human beings are born free and equal

> **Box 10.1 The essence of the human right to water**
>
> The right to water entitles everyone to have access to sufficient, safe, acceptable, physically accessible, and affordable water for personal and domestic use.
>
> **Sufficient**: the water supply for each person must be sufficient and continuous for personal and domestic uses. These uses ordinarily include drinking, personal sanitation, washing of clothes, food preparation, personal and household hygiene.
>
> **Safe**: the water required for each personal or domestic use must be safe, therefore free from micro-organisms, chemical substances, and radiological hazards that constitute a threat to a person's health. Measures of drinking-water safety are usually defined by national and/or local standards for drinking-water quality.
>
> **Acceptable**: water should be of an acceptable colour, odour, and taste for each personal or domestic use. All water facilities and services must be culturally appropriate and sensitive to gender, lifecycle, and privacy requirements.
>
> **Physically accessible**: everyone has the right to a water and sanitation service that is physically accessible within, or in the immediate vicinity of the household, educational institution, workplace, or health institution.
>
> **Affordable**: water, and water facilities and services, must be affordable for all.
>
> *Source*: UN, n.d. (a)

in dignity and rights. They are endowed with reason and conscience and should act towards one another in a spirit of brotherhood' (UN, 1948). The statement of these beliefs is one thing; their outworking more than 70 years later paints a very different picture.

Neier, for example, while strongly in support of fairer distribution of resources and of the universal application of civil and political rights, sums up what for him is the heart of the matter in regard to economic and social rights:

> whenever you get to these broad assertions of shelter or housing or other economic resources, the question becomes: what shelter, employment, security, or level of education and health care is the person entitled to? It is only possible to deal with this question through the process of negotiation and compromise. Not everybody can have everything. There have to be certain decisions and choices that are made when one comes to the question of benefits, and a court is not the place where it is possible to engage in that sort of negotiation and compromise. It is not the place where different individuals come forward and declare their interests and what they are willing to sacrifice for those interests. That is the heart of the political process; only the political process can handle those questions. (Neier, 2006: 2)

A fundamental problem, which proponents of economic and social rights are arguably attempting to circumvent rather than address head-on, is the limited reach of democratic governance and voice in many of the countries where significant numbers of people, or specific groups of people, are excluded from public services such as water supply. Quoting Neier again:

> Finally, I would say that it is important to recognize how significant civil and political rights are in dealing with economic and social inequities. Probably the best-known work on this subject is Amartya Sen's research on famine [Sen, 1981, 1987, 1999], in which he persuasively demonstrated that no famine has occurred since the end of the Second World War in countries where there was democratic accountability and the ability to communicate freely. (Neier, 2006: 3)

As part of its work to promote democratic freedoms, the United Nations General Assembly has listed the attributes of democratic governance (Box 10.2). Two observations are clear: first, these attributes are far from universal, and even in countries that purport to uphold them, violations are common; second, where the freedoms, checks, and balances included in this list are seriously deficient, it is naïve to expect genuine respect for either the rule of law or the pursuance of human rights.

Supporting national governments. In their attempts to influence and improve the enabling environment for rural water, international actors including the African Ministers Council on Water, Stockholm International Water

---

Box 10.2 Essential elements of democracy

'The values of freedom, respect for human rights and the principle of holding periodic and genuine elections by universal suffrage are essential elements of democracy. In turn, democracy provides an environment for the protection and effective realization of human rights.

For several years, the UN General Assembly and the former Commission on Human Rights endeavoured to draw on international human rights instruments to promote a common understanding of the principles and values of democracy. As a result, in 2000, the Commission recommended a series of legislative, institutional and practical measures to consolidate democracy. Moreover, in 2002, the Commission declared the following as essential elements of democracy:

- Respect for human rights and fundamental freedoms
- Freedom of association
- Freedom of expression and opinion
- Access to power and its exercise in accordance with the rule of law
- The holding of periodic free and fair elections by universal suffrage and by secret ballot as the expression of the will of the people
- A pluralistic system of political parties and organizations
- The separation of powers
- The independence of the judiciary
- Transparency and accountability in public administration
- Free, independent and pluralistic media.

Source: UN, n.d. (b)

Institute, Skat, UNICEF, and the World Bank have engaged in a variety of initiatives to advance the sector. Such approaches have included:

- country status overviews – two rounds of studies of African countries, published in 2006 and 2011, but not subsequently maintained (de Waal et al., 2011);
- bottleneck analyses – a participative process to 'enable a systematic identification of factors (or "bottlenecks") that prevent achievement of sustainable service delivery within national or sub-national WASH targets and help stakeholders to define activities aimed at removing the root causes of these bottlenecks' (UNICEF and SIWI, 2019);
- financial tracking – a means of tracking financial flows into and through the WASH sector (UN-Water and WHO, 2015);
- sustainability checks – a framework for assessing the current and likely future sustainability of services (UNICEF, 2017);
- joint sector reviews – national annual meetings between all stakeholders, during which progress is reviewed and commitments are made (Danert et al., 2016).

The overall aim of these initiatives at national level has been to encourage commitments and actions based on solid evidence and analysis.

Much has been done in the 21st century to promote codes of practice and professionalization of the rural water sector. The Rural Water Supply Network (RWSN) has produced a series of documents, beginning with and subsequently building on its 'code of practice for cost-effective boreholes' (Danert et al., 2010; see also Chapter 4). Support for the creation of professional associations of individuals and entities active in rural water provision has been part of RWSN's agenda for several years (RWSN, n.d., g). For IRC (the Netherlands-headquartered WASH organization), together with RWSN and UNICEF, the idea of professionalization of rural water services has also been prominent (see, for example, Lockwood and Le Gouais, 2015). Although the word 'professional' can mean either (a) competent or skillful, or (b) paid as opposed to voluntary or amateur), it is the second meaning that has dominated the discourse so far. I say more about this in Chapter 11.

There has to date been limited work on the topic of formal regulation of the rural water sector. However, a recent study of the subject by Gerlach (2019: 42) concluded by favouring a pragmatic approach consisting of 'performance monitoring with added advisory support offered by a trusted, professional partner (i.e. the "regulator")'. This might later evolve toward a more conventional regulatory model focused on accountability, similar to the model used in the urban water sector in many countries.

All these initiatives have been undertaken in order to strengthen the hands of national and local governments in raising and maintaining standards of work in the rural water sector.

Good governance initiatives. Beyond the water sector, but nevertheless of great relevance to it, a number of bilateral and multilateral agencies have been working for many years to promote good governance. The UK government, for example, opened a 2019 position paper on the subject thus:

> Governance is about the use of power, authority and how a country manages its affairs. It concerns the way people mediate their differences, make decisions, and enact policies that affect public life. It shapes whether people are poor or prosperous, free or oppressed. It is central to whether a young person can get a job, whether a woman can own land, or whether a life is cut short by violence. In many developing countries, governance is the dominant constraint to inclusive growth. (DFID, 2019: 4)

Transparency and accountability are at the heart of good governance, and the various attempts to reform institutions and organizations have focused especially (but not only) on these aspects. The logic of institutional and organizational reform is clear (Figure 10.3). Changing the way decisions are made and how power is used – institutional change – at the same time as changing the way things work within organizations, are means to the end of improved service delivery. The logic is clear, but the realities are often very different. This kind of systems change is far from easy in practice. As Andrews et al. (2012: abstract) describe the problem, 'Many reform initiatives in developing countries fail to achieve sustained improvements in performance because they are merely isomorphic mimicry – that is, governments and organizations pretend to reform by changing what policies or organizations look like rather than what they actually do.'

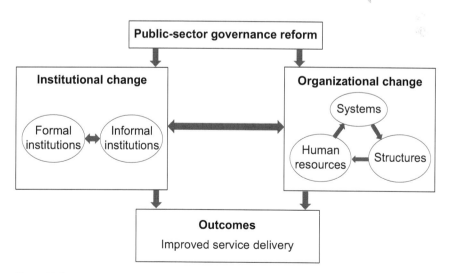

Figure 10.3 The logic of institutional reform
Source: Joshi and Carter, 2015

General governance and institutional reform initiatives are relevant to the water sector, as many of the weaknesses in governance in low-and middle-income countries affect all public services. Good decision-making regarding allocation of resources; investment in training and education of public servants; and responsiveness to demand all affect the rural water sector deeply. Work on governance and accountability in the WASH sectors aims to bring about a situation in which 'all decision-makers in government, the private sector and civil society organisations recognise that being open and transparent, engaging stakeholders, evaluating and learning, and responding to complaints is crucial to their legitimacy and effectiveness, and to achieve long-lasting benefits to the poor of sustainable water, sanitation, and hygiene interventions' (Jiménez et al., 2015: 5).

Exposing and fighting corruption. Numerous international organizations and most national governments are involved in the fight against corruption. Transparency International works with a wide range of like-minded partners, 'enabling and facilitating a culture of anti-corruption action', 'advocating for anti-corruption laws and prevention systems to be adopted', and seeking to 'grow [its] influence and innovate in [its] work' (Transparency International, 2020: 9, 12, 15). The Water Integrity Network describes water integrity as 'honest, transparent, accountable, and inclusive decision-making by water stakeholders, aiming for equity and sustainability in water management' (Water Integrity Network, n.d.). Both organizations, and others engaged in the same struggle, are working to expose corrupt practices, mobilize society to oppose them, and bring those responsible to justice in the courts. This is an uphill battle, however, like all work to bring about fundamental systemic change.

Introducing systems strengthening

Much of what I have outlined in this chapter describes activities to change and strengthen 'the system' so making it more effective. Generating evidence, strengthening accountability mechanisms, supporting local and international demands for change, and fighting corruption may all be components of a concerted effort to move from business as usual to ways of working that are better fitted to the final decade of the SDG period. I return to the matter of systemic change and systems strengthening in Chapters 11 and 12 of the book.

CHAPTER 11
What's changing in rural water supply?

Abstract: *Although there has been major global progress in the provision of rural water services during the 21st century, access to 'at least basic' service in 2017 was less than 70 per cent across 49 countries. A number of patterns emerge, including the relationship between 'at least basic' access and gross national income per capita. Some countries defy the general trends, either under-performing in relation to their economic level, or over-performing. There is still a long way to go to achieve basic, on-premises, or safely managed services for all. Various technology innovations show promise in ameliorating inadequate supply systems, the main ones being in the domain of information and communication technologies. These are beginning to enable more effective monitoring of rural water systems; however, the extent to which they will translate into better performance depends on the interface with social and institutional arrangements. In addition, the sector has also seen shifts in management trends – dominated by a desire to move beyond community management to professionalized or utilitized arrangements – and in financing trends, where payment for independently verified results and attempts to broaden and blend multiple financial streams prevail. Overall, water, sanitation, and hygiene and rural water sector thinking has been influenced by the science of systems, especially complex adaptive systems, and systems strengthening.*

Keywords: service levels, technology, development, monitoring, water management, financing, systems thinking, systems strengthening

> 'You must be the change you wish to see in the world'
> —Mahatma Gandhi, 1869–1948

Introduction

In this penultimate chapter I explore what is changing in rural water supply and how these trends are occurring across regions and nations. The rural water sector is dynamic in many ways – in regard to technology, monitoring, levels of service, models of management, and approaches to financing. The over-arching thinking, which both reflects these evolutionary changes and which at the same time helps to drive change, is evolving too. Innovations in technologies and approaches that are introduced either fade into oblivion (if they fail to gain traction), or become the accepted way of doing things (if they succeed). Consequently, the list I highlight here will inevitably become dated.

In this chapter I show first what has been changing so far during the 21st century, both globally and for the least well served, in regard to rural water services. I then move on to consider how innovations and new approaches in technology, monitoring, management, financing, and overall water sector thinking are attempting to accelerate progress in the sector.

Levels of service

The best quantitative data regarding access and level of service is that assembled by the Joint Monitoring Programme (JMP) from national household surveys. The JMP distinguishes five levels of service, culminating in so-called 'safely managed' water supply in which water is provided on-premises, available at least 12 hours out of every 24, and free from contamination. Table 11.1 sets out brief descriptions of the service levels, and how absolute numbers of rural people served at each level have changed between 2000–17 (the latest year for which data was available at the time of writing). Figure 11.1 shows these data visually. Data were available for 167 countries, representing 99 per cent of global rural population.

Table 11.1 Service levels and change, 2000–17, rural water supply (global)

Service level	Description (JMP, 2017)	Numbers served, 2000	Numbers served, 2017
Safely managed	Piped or non-piped improved drinking water in dwelling, yard, or plot, available when needed, *and* free of contamination	1,266,926,879 (39%)	2,013,734,409 (60%)
Basic	Drinking water from an improved source, no more than 30 minutes round trip including queueing	945,014,393 (29%)	716,150,584 (21%)
Limited	Drinking water from an improved source, more than 30 minutes round trip including queueing	127,833,827 (4%)	153,015,401 (5%)
Unimproved	Drinking water from an unprotected well or spring	632,509,038 (20%)	359,933,824 (11%)
Surface water	Drinking water from a river, dam, lake, pond, stream, canal, or irrigation canal	237,897,226 (7%)	127,246,481 (4%)
Totals		3,210,271,362	3,370,080,699

Note: Statistics for 'safely managed' in this table are only for on-premises component of that category, as too few countries are yet providing full data for availability and water quality; improved sources include piped water, boreholes, protected dug wells, protected springs, rainwater, packaged water, delivered water
Source: JMP, 2020b

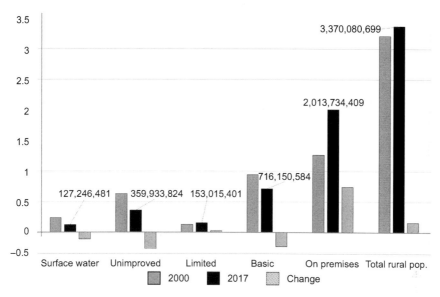

Figure 11.1 Service levels and change, 2000–17, rural water supply, global
Source: JMP, 2020b

Table 11.1 and Figure 11.1 show, from a global perspective, the following changes between 2000 and 2017:

- Rural population rose over the period by (only) 5%.
- The lowest two service levels (surface water and unimproved) saw reductions in absolute numbers of people thereby served by nearly half in aggregate.
- Among the three 'improved' levels of service (limited, basic, and on-premises), numbers of people served at limited and on-premises level both rose (by 20% and 59% respectively), while the numbers with a basic service fell (by 24%).
- Overall, the numbers of people enjoying improved water service levels rose by about one quarter over the period.

Although the global picture is generally encouraging, global trends conceal regional and national differences. Furthermore, 487 million rural people were still using surface water or unimproved sources in 2017, and if limited service is added in, this figure rises to 640 million.

Ranking countries according to the proportions of their rural populations enjoying at least basic service (i.e. the total of basic and on premises), and examining the changes since year 2000 is revealing. Table 11.2 lists the bottom 49 countries ranked in this way (I have chosen that figure arbitrarily as it includes all countries having less than 70 per cent of their rural populations with access to at least basic service in 2017). Figure 11.2 shows these statistics visually.

Table 11.2 Bottom 49 countries ranked by 'at least basic' (sum of basic and on premises) in 2017 (percentage access by rural population)

Country	Basic, 2000	Basic, 2017	On premises, 2000	On premises, 2017	Total, 2017
Timor-Leste	19.5	14.7	23.8	55.0	69.7
Namibia	42.9	23.6	23.6	45.6	69.2
Mali	27.6	50.9	10.2	17.4	68.3
Ghana	53.8	56.0	0.2	11.5	67.5
Malawi	45.5	57.7	1.8	7.7	65.4
Gambia	64.1	55.5	3.7	7.9	63.4
Liberia	43.3	57.1	6.2	4.9	62.0
Solomon Islands	19.9	18.6	56.6	42.0	60.6
Eswatini	26.5	24.3	16.0	36.1	60.4
Lesotho	63.1	53.0	1.3	6.3	59.3
Nicaragua	37.2	5.1	26.2	54.0	59.1
Benin	28.7	46.2	23.3	12.0	58.2
Côte d'Ivoire	39.7	40.1	15.9	17.7	57.8
Afghanistan	11.9	32.6	9.7	24.7	57.3
Burundi	46.9	53.8	0.9	2.8	56.6
Mongolia	27.6	47.5	1.2	8.4	55.9
Nigeria	23.8	39.9	11.1	15.7	55.6
Gabon	29.3	31.9	7.7	23.2	55.1
Yemen	9.4	20.0	16.2	34.7	54.7
Guinea-Bissau	37.5	41.4	3.9	12.1	53.5
Sudan	18.6	28.1	16.7	25.1	53.2
Rwanda	40.6	48.3	0.4	4.3	52.6
Sierra Leone	22.9	41.8	1.5	8.3	50.1
Mauritania	9.7	21.2	15.3	28.7	49.9
Zimbabwe	43.6	35.6	16.3	14.2	49.8
Kenya	22.1	34.4	15.1	15.2	49.6
Guinea	53.1	28.6	0.0	20.1	48.7
Togo	25.2	42.5	4.4	5.9	48.4
Djibouti	49.5	42.6	5.2	4.5	47.1
Congo	11.1	26.6	6.9	19.1	45.7
Niger	22.1	34.9	2.3	8.7	43.6
Haiti	30.6	34.9	9.9	7.7	42.6
Tanzania	16.4	26.5	0.0	16	42.5
Zambia	26.7	35.6	4.2	6.4	42.0
Uganda	18.1	36.8	1.1	4.5	41.3
Mozambique	4.1	33.2	0.0	6.8	40.0

Country	Basic, 2000	Basic, 2017	On premises, 2000	On premises, 2017	Total, 2017
Cameroon	34.1	32.9	3.8	6.1	39.0
Madagascar	22.4	20.8	1.4	15.5	36.3
Burkina Faso	50.8	31.6	0.0	3.4	35.0
South Sudan	37.7	32.8	1.0	2.1	34.9
Papua New Guinea	12.3	16.0	14.2	18.6	34.6
Central African Rep.	40.7	30.9	3.3	2.8	33.7
Ethiopia	8.6	26.5	0.0	4.6	31.1
Equatorial Guinea	36.2	28.8	2.5	2.0	30.8
Chad	28.3	28.2	3.4	1.3	29.5
Somalia		19.9	0.3	8.2	28.1
Eritrea	21.6	19.4	9.4	8.4	27.8
Angola	19.4	20.2	1.8	7.2	27.4
DRC		22.3	0.7	0.5	22.8

Note: Earliest data for Timor-Leste are for 2002, and for South Sudan 2011
Source: JMP, 2020b

The changes in Table 11.2 place individual countries into four categories:

- Those where access to basic service has increased significantly, and on premises service has also increased (usually by a smaller percentage): Mali, Malawi, Afghanistan, Burundi, Mongolia, Nigeria, Yemen, Sudan, Rwanda, Sierra Leone, Mauritania, Togo, Congo, Niger, Tanzania, Zambia, Uganda, Mozambique, Papua New Guinea, Ethiopia.
- Those where basic level of service has stayed approximately level, or fallen, but on premises service has increased (in some cases by an impressive amount): Timor-Leste, Namibia, Ghana, Gambia, Eswatini, Lesotho, Nicaragua, Gabon, Guinea Bissau, Guinea, Cameroon, Madagascar, Burkina Faso, South Sudan, Somalia, Angola.
- Those where access to basic service has increased, but on premises access has not increased: Liberia, Benin, Kenya, Haiti.
- Those where access at both service levels has stayed roughly the same or fallen: Solomon Islands, Côte d'Ivoire, Zimbabwe, Djibouti, Central African Republic, Equatorial Guinea, Chad, Eritrea, Democratic Republic of the Congo (DRC).

Among the bottom 49 countries, access to basic services (in terms of percentage of rural population) is on average more than double that of on premises access.

An important determinant of service level and access is the state of the national economy. Out of the 49 countries in Table 11.2, 27 (55 per cent) are

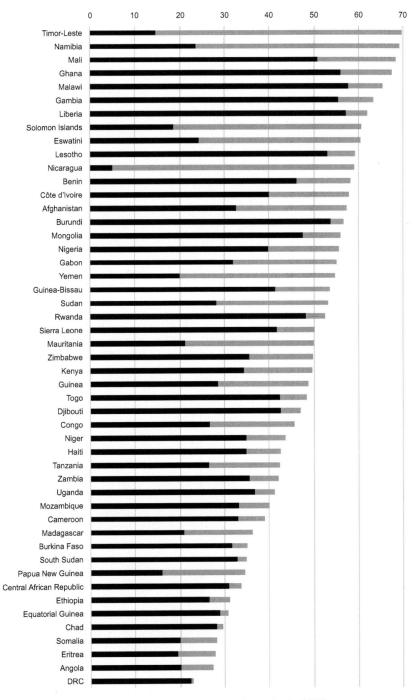

Figure 11.2 Bottom 49 countries ranked by 'at least basic' service in 2017
Note: Left portion of each bar (in black) is percentage access to basic service; right portion (in grey) is percentage access to on-premises service. Totals are access percentages to at least basic service. All statistics are for rural populations only

WHAT'S CHANGING IN RURAL WATER SUPPLY?

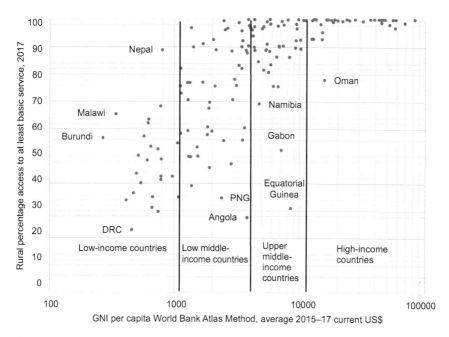

Figure 11.3 Gross national income per capita vs rural water 'at least basic' access (%), 2017
Source: JMP, 2020b; World Bank databank, 2020d

low-income and 19 (39 per cent) are lower middle-income countries (a total of 94 per cent). All but two of all the world's low-income countries are in this list.

Figure 11.3 shows the relation between gross national income (and World Bank income group) and rural access to at least basic service in 2017, for all 167 countries for which JMP has rural access estimates. A few countries are specifically identified on the figure. The broad relationship between national income and access to rural water is clear, but there is a good deal of scatter, and a number of notable outliers, around the general trends.

In general, it is evident that:

- at least basic access by more than 90% of national rural populations only becomes commonplace when countries enter the upper middle-income or high-income categories;
- low- and lower middle-income countries struggle to provide at least basic services, with more than 20 countries having less than 50% access to this level of service;
- some countries defy the general trends, either having relatively high national incomes but low levels of service (e.g. Oman, Equatorial Guinea, Gabon, Namibia, Angola, Papua New Guinea); or having low national incomes but relatively high levels of access to at least basic service (e.g. Burundi, Malawi, Nepal).

Those countries that appear to be underperforming in relation to the state of their economies include some facing natural challenges, others dealing with conflict, and still others in which government budget allocations are simply insufficient to address their people's needs. Conversely, those that are apparently over-performing in relation to their national income either likely place greater policy priority on rural water services, or enjoy significant contributions from foreign donors and non-governmental organizations (NGOs).

A general comment on these statistics and trends: while global totals provide some cause for celebration, the performance of individual countries is very variable. Regional generalizations are not particularly helpful, since each geographic region contains some well performing and some poorly performing nations. The World Bank's country classification by gross national income (GNI) is informative in highlighting the relationship between national wealth and the level of public services provided to its citizens. Other country categorizations – for example by 'fragility' or other indicators of governance and democratic accountability – would also reveal informative patterns.

Finally, as Table 11.1 shows, about 1.35 billion rural people still do not have water supplied on the premises; and this is only one of the three requirements of a safely managed service. The rural water sector does indeed have a long way still to go (Roche et al., 2017).

Technology

Technology is generally understood to include both the hardware and software that human beings use in almost every area of life, and the science, craft, and skills needed to produce and maintain them. Although there are huge disparities in access to both the range and sophistication of technologies, probably everyone on the planet uses and is affected by technology.

In terms of technology in the water sector, I focus first on technologies that form part of the physical infrastructure of water delivery, and then on those that help to observe and monitor how that infrastructure is performing.

A recent global evaluation of UNICEF's work in rural and small-town water supply only identified one technology innovation – real-time monitoring – in the work of the largest international organization that is focused on rural water supply (working in about 80 countries worldwide) (UNICEF, 2018). Here I highlight a few more, while recognizing that innovations in areas other than technology may be more pertinent to the sector.

Table 11.3 sets out some recent innovations, with a brief commentary on each. It is perhaps unsurprising that technology innovation has tended to move beyond traditional areas of drilling technology, pump design, water storage, treatment, and delivery, to applications using information and communications technology, so-called internet-of-things, and cloud computing (Andres et al., 2018; Thomas et al., 2018). Nevertheless, the table includes some of the former category, too.

Table 11.3 Recent technology developments in rural water supply

Innovation	Reference	Comments
Digital tools for drilling	Practica Foundation (the Netherlands) https://www.practica.org/wp-content/uploads/2020/02/DTD_infosheet_Practica.pdf	Three mobile phone apps to allow field analysis of vertical electrical sounding (geophysics) data, drill log generation and pumping test data entry, and reporting
Digital data management systems for water points	Madzi Alipo (originated in Malawi) https://www.madzialipoapp.org MWater https://www.mwater.co Open Data Kit (ODK) https://opendatakit.org	Systems for collecting, storing, uploading, displaying, and analysing water point data. Madzi Alipo and mWater are specific water sector apps, while ODK is more generic
Water quality testing	Sorensen et al. (2015) Nowicki et al. (2019)	Use of tryptophan-like fluorescence to detect faecal contamination of water
Real-time SMS reporting	UNICEF (2018) use of U-report and Rapid Pro text messaging tools https://www.unicef.org/innovation/U-Report https://www.unicef.org/innovation/rapidpro	Enabling citizens to receive and supply information
Sensors on components of water physical infrastructure	Smart handpump project (University of Oxford) https://www.ox.ac.uk/research/research-impact/smart-handpumps Charity: water handpump sensor programme https://www.charitywater.org/our-work/sensors SweetSense 'internet-of-things' application http://www.sweetsensors.com http://www.sweetsensors.com/our-technology/	Various types of sensors and related technology for monitoring handpump, electric submersible pump, and piped system performance Sensors variously communicate via mobile phone networks or to satellite, storing data in the cloud
Deep-well handpumps and corrosion-free rising mains	Collaboration between WaterAid and Poldaw Lifepump https://lifepump.org See also Chapter 5	Working to extend handpump range while addressing corrosion and strength issues in below-ground pump components
Low-cost reliable solar water pumps	Impact pump (Thermofluidics Ltd, UK) https://www.impactpumps.com See also Chapter 6	Focus on low price and high reliability of below-ground pumping and delivery system
Standalone water vending systems	One of many: Grundfos Lifelink https://www.grundfos.com/market-areas/water/lifelink.html See also Chapter 6	Usually consisting of a solar-powered pump delivering water to an overhead reservoir, with treatment to potable quality, and vending by means of tokens or mobile money

176 RURAL COMMUNITY WATER SUPPLY

What is apparent from Table 11.3 is that technology innovations mostly fall into three broad categories:

- Those that enable remote monitoring of functioning of entire water supply systems, or components of them. These provide the possibilities for water service managers to (a) know better what is the status of the water supply systems under their responsibility, and (b) to act accordingly, in an efficient manner.
- Those that enable easier, more reliable, or lower cost access to groundwater. In various ways these provide the possibility of extending the mean times to failure of pumping systems, and reducing downtimes.
- Those that enable easier real-time or near-real-time visualization and reporting of standard procedures in the development of groundwater – borehole siting, drilling reporting, and test pumping reporting.

All these, and the entire list in Table 11.3, provide highly desirable but not sufficient conditions for progress. Technology only advances human progress to the extent that it is used, used properly, and managed and financed effectively.

Approaches to monitoring

National progress, tracked internationally

International monitoring of progress towards rural water goals as part of wider water, sanitation, and hygiene (WASH) sector progress monitoring is well described by Bartram et al. (2014). One of the most important changes to the way monitoring of coverage (access) is undertaken is from the use of government provided data – which often simply multiplied cumulative numbers of taps and handpumps (for example) by notional numbers of users – to the use of household survey data. This change took place at the end of the 20th century and was implemented in the JMP report of 2000. Unsurprisingly, household survey and government-provided data often diverged quite considerably because of the different methods used in their estimations. Since 2000 JMP only reports access and related statistics on the basis of household and institutional (health care facilities and schools) survey data (JMP, 2019a).

The Joint Monitoring Programme has always focused on coverage, and the numbers and proportions of populations that have access to services at different levels (the latter disaggregated by rural and urban habitation, water and sanitation, and more recently hygiene access and disaggregation by wealth quintiles).

In 2008 a complementary monitoring initiative was introduced, examining and tracking the key obstacles and enablers of progress nationally, notably governance (by which is meant legislation, policies, plans, and regulatory frameworks), institutional arrangements, financing and financial systems, monitoring systems, and human resources. This is the UN-Water Global

Analysis and Assessment of Sanitation and Drinking Water (GLAAS) (UN-Water, 2019). In a sense GLAAS monitors the inputs or means, and JMP the outcomes or ends, of efforts to extend water and sanitation access.

Local level and national monitoring

Progress at local government level and that achieved by the efforts of NGOs should, in principle, be reported upwards to national government. Local government reporting is one of the key functions undertaken by district water officials and administrators. Unfortunately, NGOs are not always compliant with requirements to report their achievements in a timely manner to the local government administrations within which they operate – although the importance of doing so is increasingly recognized by responsible actors in the sector.

One of the main constraints experienced by local governments in regard to monitoring rural water supply access and performance is their limited resources, both human and financial. Too few staff, too few vehicles, and insufficient funds to run the vehicles that they do have, means that regular visits to remote communities are simply not possible. Real-time monitoring using SMS reporting by communities, or sensors installed at water points (Table 11.3), can make a real difference here, since physical visits can be reduced in number to those pinpointed by such systems.

Monitoring access, sustainability, and service levels

An important gradual trend in monitoring has been from simply counting the numbers of people assumed to have access, to an increasing focus on the level of service and the performance of services that people actually enjoy; this also includes assessing the performance of the providers of those services. Monitoring whether or not water points are working (functionality) provides a snapshot that, if repeated from time to time, can indicate the direction of progress (or failure to improve). However, much remains to be done to improve understanding (and trigger action) of service level and performance. I would argue that sets of indicators much closer to those used in urban water utility performance monitoring need to be developed. Table 11.4 shows the relevance of such indicators (bearing in mind that the utility-provided services assumed in the International Benchmarking Network (IBNET) are exclusively via piped supply).

The idea of rural water supply indicators being modelled on those generally accepted for utilities is not original. Building on the earlier work of the International Water Association (Alegre et al., 2016), Haider et al. (2014) reviewed the various frameworks of performance indicators for small and large water utilities, proposing a step-wise approach to the adoption of appropriate indicators. A subsequent paper by the World Bank (2017b) proposed 24 indicators of service level, functionality and sustainability at

Table 11.4 Water utility indicators

IBNET indicator	Brief description	Relevance to rural water supply in low-income countries
Service coverage	Percentage of population with access, disaggregated by household and public connections	An additional disaggregation to account for non-networked water points is needed
Quality of service	Measures (a) continuity of supply, (b) water testing and quality, and (c) number of complaints	All are highly relevant, with modifications at level of detail only
Water consumption and production	Water produced and water consumed per connection, household or individual as appropriate	Less relevant for point sources, but needed for piped systems
Billing and collections	Revenues and effectiveness of billing	The relationship of revenues collected to actual operations and maintenance costs is critical
Non-revenue water	Measures of water 'lost' to the system by leakage, theft, or use without payment	Less relevant with point sources (except to the extent that e.g. handpumps are used without payment); highly relevant in piped systems
Financial performance	Ratio of total annual operational revenues to total annual operating costs, and ability to repay any debt	Highly relevant to all service levels
Metering practices	Number of connections and volume of water metered (as percentage of totals)	Applicable in piped systems if costs of metering are not disproportionate
Assets	Value of fixed assets per person served	Relevant, but alongside consideration of minor opex and capmanex costs
Pipe network performance	Number of pipe breaks per year per km of pipe	Relevant in piped systems. Equivalent with point sources: mean time between failures and time to repair
Affordability of services	Annual water bill for use of 6 m^3 per household per month Annual operating revenues per capita divided by annual per capita GNI	Relevant when compared to actual cash income of rural households
Cost and staffing	Operational costs per m^3, staff per 1000 connections, labour and energy costs	Relevant in private operator (piped) systems
Process indicators	Planning, staff training and management, service level offerings, customer complaints system	Very relevant in all cases

Source: IBNET, n.d.

Table 11.5 Suggested rural water supply performance indicators

Indicator group and proposed indicators	
Service level	(13) Chlorination
(1) Type of source	(14) Coverage
(2) Accessibility	(15) Non-revenue water
(3) Availability	**Financial management**
(4) Quality	(16) Tariff structure
(5) Reliability	(17) Financial management
(6) Affordability	(18) Tariff collection efficiency
(7) User satisfaction	(19) Financial sustainability
Functionality	**Environmental and water resource management**
(8) At level of individual handpump	
(9) Physical condition of the water supply infrastructure	(20) Source, catchment, and water resources management
Governance	**Customer relations**
(10) Presence of a legally established service provider	(21) Complaints handling mechanism
	Service authority presence and functions
(11) Staffing	(22) Service authority capacity
Performance in operation and maintenance	(23) Service authority support functions
(12) Maintenance	(24) Presence of an information system

Source: World Bank, 2017b; for detail see Annex Tables A.1–A.4 of that report

minimum, basic and advanced level for rural water systems, and suggested a way forward for validating and adopting such measures (summarized here as Table 11.5).

In view of efforts to professionalize the rural water sector (whether in the sense of simply enhancing competence of those responsible, paid or not, or introducing monetary recompense), it would seem desirable that rural water sectors should move beyond single-indicator performance measures such as functionality to more comprehensive frameworks similar to those just described.

In closing this section, it is important to highlight the common gap between monitoring – the collection of data – and action to respond to the information, which those data can provide (WaterAid, 2020). Without such action, monitoring is a somewhat futile process.

Management and financing

Management trends

Two broad trends in thinking about the management of rural water services are evident. The first is a tendency to reject the dominant model that has

> **Box 11.1 Some weaknesses of community management**
>
> Community management of rural water supply, as applied since the 1980s, depends, among other things, on the following attributes:
>
> - It relies on **voluntary service** by elected community members, who take responsibility for management, maintenance, fund-raising for repairs, and arranging for or undertaking repairs.
> - It assumes that a short period of **training** of committee members at the time of installation of a new water point is sufficient to transfer necessary technical, managerial, and financial skills to the committee.
> - It relies on a level of **trust** within the community and its representatives (usually a water point or water management committee) over finance and other aspects of accountability.
> - It assumes that **conflicts** or disagreements within the community can be effectively resolved by the committee.
> - It assumes that **succession** from one set of committee members to the next generation can take place readily without external help.
>
> In reality, voluntarism can work for a time, but volunteers eventually tend to lose interest or find other priorities demanding their attention. The management of even apparently simple technology such as a handpump requires varied and unfamiliar skills, which cannot be transferred in a few days of training – especially in the absence of refresher training and updating. Trust, especially over financial matters, is often weak or absent within rural communities and between those communities and external organizations. Conflict resolution is inherently challenging in the absence of fully recognized authority and respect. Succession and the transfer of expertise to future generations does not happen automatically.

served the sector since the 1980s – community management. All who work in the rural water sector are aware of the many weaknesses and flaws of the model (Box 11.1), but the debate nevertheless continues as to whether the model should be abandoned or made to work by more effective implementation.

The second trend is characterized not by the rejection of an old model, but by the promotion of ideas such as professionalization (in the sense of services run by paid personnel) and utilitization. In a 2019 paper, Franceys argues that because the internationally driven imperative to raise service levels and access have been so far ahead of the true demands of rural communities that would naturally develop by concurrent social, cultural, economic, and institutional development, only two alternative solutions exist. One is for governments to heavily subsidize rural water services – not only the capital costs as at present, but also operating and capital maintenance costs. The other solution is what Franceys calls 'utilitization' – the expansion of urban and small town water utilities into their surrounding rural hinterlands, with concurrent development of 'micro-utilities' further out in the rural areas. These micro-utilities would eventually be swallowed up by the urban utilities, or they would coalesce to enable economies of scale. Franceys' argument is based on the fact that urban water utilities – despite their own weaknesses – are 'reflective of (co-evolved with) governance capacity in the setting in which they have to operate.

However imperfect, they are society's designated organisation to integrate the service and financial challenges of delivering improved water and sanitation to all in urban areas' (2019: 6).

Financing trends

Until recently, the landscape of rural water financing has been fairly simple. The capital investment to extend access has come from national budgets disbursed through central and local government; donor, international non-governmental organizations, and philanthropic grants to governments, UN agencies, and local NGOs; and loans to national governments from the development banks. These funding organizations (national and foreign) have focused almost entirely on capital expenditure, and budget allocations to support operation, maintenance, and capital maintenance and replacement have been the exception rather than the rule.

Two streams of thought have developed with regards to improving financing models. First is the recognition that without an explicit focus on post-construction (recurrent) expenditure, it is impossible to achieve sustainability in rural water supply. The second trend is a move towards so-called results-based financing, in other words retrospective payment by funders for certain defined outcomes, rather than up-front grants to cover the costs of implementing projects and programmes. These two trends converge, but first I unpack each separately.

The learning initiative 'Sustainable Services at Scale' (known as Triple-S) was funded by the Bill and Melinda Gates Foundation and undertaken by IRC between 2009 and 2014. It built on earlier WASH sector ideas about services and sustainability, and aimed to influence sector thinking to move away from simply providing new infrastructure to providing lasting services. Overall, the project was effective at 'putting sustainable services on the map' in global debates, but less so at bringing about specific changes on the ground in its focus countries (Hydroconseil/Trémolet Consulting, 2015). Nevertheless, its legacy continues, especially through its combination with the thinking generated in the concurrent WASHCost project, and subsequent thinking about systems and systems strengthening. The end-of-project evaluation was, however, arguably undertaken too soon after the event to give this longer-range perspective.

One outcome of the work undertaken both by Triple-S and WASHCost was the increased focus on the operation, maintenance, repair, and rehabilitation of rural water infrastructure, and of the real costs and affordability of those post-construction aspects of rural water services. It has become increasingly apparent to most professionals in the sector that increasing access is not enough, in the absence of attention to sustainability, and that the question of who pays (and who can afford to pay) the recurrent costs has to be addressed with rigour.

Those organizations funding international development work, including rural water, are increasingly interested in results and impact. Donors wish to see their money spent well, and this means reducing the incidence of failures and increasing beneficial outcomes for the recipients of aid funding. Some donors also want to transfer the risks inherent in all development programmes to those implementing them – so meaning that their money is only spent on success, leaving implementing organizations to pick up the tab for failed interventions. These two trends in donor thinking combine in so-called 'results-based financing' or 'output-based aid'.

In the WASH sector, one of the largest initiatives using this financing model was the UK Government's WASH Results Programme, 2013–21, implemented in 11 countries of sub-Saharan Africa and South Asia, which referred to its modality as 'Payment by Results' (PBR) (Development Tracker, 2020). This kind of funding modality involves a contract between the donor and the implementer to deliver pre-defined 'results', ideally at outcome level in logframe terms (not simply numbers of taps and toilets); the delivery of those results is verified to the donor by an independent assessor; then, after verification, the implementer is paid.

Payment by results has certain attractions – especially for donors – but it also has drawbacks. Smaller implementing organizations cannot easily pre-finance their work, especially when they are set up to receive grants, by tranches, in advance of their activities. But the fundamental question is who should take the risk for delivering inherently risky work? Not all development interventions work as planned; some failure is inevitable. Who should pay for this? In the past, grant funding largely loaded that risk on the donor. Payment by results takes things to the opposite extreme, loading all the risk on the organizations that are arguably least able to bear it. In Chapter 4 I gave the example of no-water-no-pay contracts for well drilling, a specific instance of PBR financing, which often leads to perverse incentives to cheat and increased real costs. I argue that risk, in a risky endeavour, should be shared by all those playing a part in the intervention.

Furthermore, in a PBR modality, the costs of independent verification of results can be high, especially if those results are not defined with great care. Despite the real practical difficulties of implementing PBR approaches well, they continue to be attractive to those seeking new ways of funding the sector (McNicholl et al., 2020). My view is that they should be adopted only with extreme caution and awareness of their possible unintended impacts.

Awareness of the immense costs of achieving the Sustainable Development Goals (SDGs) has led those thinking about development financing to explore the idea of 'blended finance'. By this is meant primarily the addition of commercial sources of finance into the existing mix of official development assistance and philanthropic/charitable finance (OECD, 2018). It seems unlikely, however, that commercial finance can be drawn into a sector such as rural water supply, which is unlikely to provide financial returns in the foreseeable future. In general, rural water supply is a public service, not a profit-making commercial opportunity.

Finally, the idea of non-grant funding for organizations (especially social enterprises) implementing development programming is being considered by various donors. The aim here is to provide funding on a repayable loan or equity basis, in order to keep money circulating and being reinvested rather than simply spent once. Again, time will tell whether such ideas can gain traction in rural water services.

How sector thinking has been evolving

During the first international water decade of the 1980s, much of the work focused on extending access to physical infrastructure. However, the limitations of this approach were fast becoming apparent, resulting in an increasing emphasis on the software of community participation and management – not that this was always done well. The notion of rural water supply as a socio-technical endeavour has its roots in this period. In subsequent decades, thinking has evolved from infrastructure to services, and from services to systems.

The idea of water supply as a service has moved the emphasis from one-off provision of physical and 'soft' infrastructure (the technology and community engagement/management aspects, respectively) to the idea that water should be available when required, indefinitely – the sustainability dimension. In the 21st century, a new idea has emerged from the experiences of, and reflections on, the previous decades of effort. Complexity science has a long history, but it became increasingly prominent in the 1970s and 1980s. Some of the earliest applications of ideas about complexity and complex adaptive systems (Box 11.2) in relation to international development started to emerge in the early years of the 21st century (Rihani and Geyer, 2001; Ramalingham and Jones, 2008). Since then, such ideas have caught the imagination of academics and other thinkers working in or around international development. The application of complex adaptive systems thinking in the water sector has also been documented recently (Neely, 2019).

The main direct application of complex adaptive systems thinking lies in how working with complex (as opposed to complicated) systems should proceed. In short, if the system is merely complicated, then tried and tested rules, recipes, and checklists can be used to good effect. Follow the rules, and all will be well. Construct a project or programme logical framework, and follow it to its logical, linear, deterministic conclusion.

If, on the other hand, the system is truly complex, the only viable way forward is to proceed iteratively – making small changes, observing or monitoring their effects, recording that learning, making further change, and so on, in a series of cycles of interventions, learnings, and adaptations of future interventions. It is also possible, however, that the system is insufficiently understood but nonetheless still complicated rather than complex. Finding the answer would require the same cycle of iterative steps, thus it is my view that the distinction may have more of an academic than a practical value.

> **Box 11.2 Systems – complicated and complex**
>
> Many aspects of our world are characterized by interconnected component parts that work together. On a very limited scale, think of a bicycle with its frame, wheels, pedals, and brakes; or a car with its electrical and fuel systems, its chassis, and other appurtenances. Or think of a family – older and younger generations, interacting in a harmonious (or sometimes dysfunctional) – way.
>
> An assemblage of component parts – and their interactions – are often described as systems. Systems theory is about the concepts and principles that apply to systems in general, and the way their structure influences their behaviour and responses to external stimuli.
>
> An important distinction is made between systems that are merely complicated, and those that exhibit features of true complexity – so-called complex adaptive systems. Complicated systems are those in which the rules determining system behaviour are understood (or can be understood) and straightforward cause-effect relationships apply.
>
> Complex adaptive systems are those in which direct cause-effect relationships do not apply (or cannot be identified); some behaviours and expressions of order emerge, apparently spontaneously and in the absence of an organizing authority; and change or evolution is evident. Complexity science describes such systems as non-linear, exhibiting emergence and order, and adaptive.
>
> An example of a complicated system is the set of engineered components, the forces acting on them, and the scientific laws describing them, which underlie space travel and mankind's ability to put satellites into orbit or land a man or woman on the moon.
>
> Complex systems include natural ecosystems, towns and cities, economies, human and animal social networks, and the climate.
>
> Distinguishing complicated from complex systems is important in terms of how we should work to bring about change. However, it is not always straightforward to make this distinction; it may be that human ignorance of the internal rules of a system makes it appear complex, when it is in fact merely complicated but insufficiently understood.
>
> Some recent thinking proposes that rural water supply in low-income countries is a complex (rather than merely complicated) system.

In rural water supply, programmes need to progress in small steps, with much learning and adaptation, rather than following elaborate pre-ordained plans. Iterative learning and adaptive processes are radically different from conventional linear and mechanistic planning approaches involving pre-determined projects, programmes, planning, and budgeting cycles. Far greater flexibility is needed, in a manner that ministries of finance and other donors find uncomfortable.

Today, many look on the WASH sectors of low-and middle-income countries as poorly performing systems that need reforming and strengthening. The WASH systems strengthening approach is examined more closely in the final chapter of this book, as I attempt to set an agenda for the last decade of the SDG era and beyond.

CHAPTER 12
Imagine another world

Abstract: *Those advocating for increased investment in rural water generally appeal to justifications based on either economic or human rights based arguments. However, decision-making is based not only on data and evidence, but also on the values and experiences of those with power. Central to the success of organizations in both public and private sectors are the shared values of personnel. I propose five core values as critical in this regard: the equal value of all people; the fundamental importance of sustainable service; the mandate of national governments; the crucial place of professional standards; and the need for continual reflection and learning. Building from these, I propose a phased approach to national rural water supply progress in which priority is given to moving households away from dependence on unprotected water sources, expending increasing sums on sustainability measures (monitoring, management, and recurrent financing), and ultimately moving all households and communities up the ladder of service. In pursuing such an approach, I emphasize the importance of appropriate national water policies, adequate national budget allocations, true fiscal decentralization, compliance and coordination of all players with government policies, and the key place of local governments. Identifying weak points in the national system, and working collaboratively to address these – the systems strengthening approach – has many merits; but in future, much of the systems change may have to be internally driven and less reliant than today on external pressures.*

Keywords: investment, economics, human rights, values, phasing, water policy, systems, systems strengthening, complexity

> 'When you can imagine, you begin to create; and when you begin to create, you realize that you can create a world that you prefer to live in, rather than a world that you're suffering in'
> —Ben Okri, 2011

Introduction

As the author of this book, I had to imagine its coming into being and its completion as I planned and drafted it. Now that it is complete, I imagine that you, dear reader, share with me the desire to see a world in which all have enough. Enough of all the necessities of life – clean air, nutritious food, habitable shelter, worthwhile work, fulfilling relationships among friends, family, and community, the means of exchange, and, of course, water. In this final chapter I invite you to imagine one admittedly narrow aspect of that world, in which all rural people have enough domestic water, close to home, safe to consume, always available, and within their means, both in financial and management terms. A small but significant dream.

First, I explore how that vision or imagining of a different world can become the shared determination, not only of those of us already working to achieve it, but also of our politicians and leaders who still need to be fully convinced of its desirability. Then I briefly examine the required vision and values of organizations, which are mandated (or which adopt the mandate) to pursue the goal of sustainable rural water for all.

This book is published in the year that marks the final decade of the Sustainable Development Goals (SDGs). What then should be the specific goals or imaginings for the remaining period to 2030? I examine these goals and address how we might tackle them. I then set out a number of practical principles for those working towards our imagined better world.

Before concluding the chapter and the book, I revisit the new(-ish) thinking in the rural water sector about systems and systems strengthening. I ask whether and under what conditions the system is truly complex, or simply complicated. I also ask whether it is helpful to describe the undeniably difficult endeavour that is sustainable rural water supply as a 'wicked problem' (Rittel and Webber 1973). Finally, I return to the importance of imagination and values in the shared endeavour of sustainable rural water services for all.

Justifying investment in rural water

Why should governments, especially of lower-income countries, allocate significant proportions of their meagre national budgets to rural water? Why should foreign donors and lenders, international non-governmental organizations (INGOs), and individuals invest in this work? How should investment in rural water be justified to these and others who control the needed financial resources? When investment in rural water has to compete for attention with other sectors including health, education, energy, agriculture, communications, law and order, and many others, how can its voice be heard and listened to? Two main strands of reasoning have dominated the ways in which organizations advocating for greater investment in rural water have approached the issue. These are the economic and the human rights arguments.

The economic justification for investment is based on a simple logic, namely that a healthier society in which less time and energy need to be devoted to fetching and carrying water is a more productive society. Less sickness means less expenditure on already burdened health services; less time spent carrying water means more opportunity for productive activities.

I examined Hutton's work on the economics of water, sanitation, and hygiene (WASH) in Chapter 8. Hutton (2012a) showed that the economic benefits of adequate water supply exceed the costs, on a global basis and in most (but not all) geographic regions. The benefits considered (health and time-saving) represent only the more easily monetizable items on a long list of tangible and intangible improvements to people's lives. In short, investment in water services represents a good use of money for national governments, although of course investments in other sectors may, under specific circumstances, yield higher returns. Although economic arguments are important, they are not necessarily sufficient to persuade politicians and national leaders to make appropriate budgetary allocations. There are many economic sectors in which investments are more visible to the public and the media, and beneficial to the careers and reputations of politicians, than rural water supply.

The human rights arguments have attempted to focus not only on what their proponents would argue is right and fair (arguably the main public perception of the notion of rights), but also on justiciability – in other words the possibility of human rights claims being decided in courts of law. The argument here is that if a nation state has signed up to the International Covenant on Economic, Social and Cultural Rights (ICESCR) – and as of July 2020, 171 countries had both signed and ratified it – then it is required by international law to 'take steps...to the maximum of its available resources, with a view to achieving progressively the full realization of the rights recognized in the present Covenant by all appropriate means, including particularly the adoption of legislative measures' (OHCHR, n.d., part II, article 2.1) . The rights (plural) to water and sanitation, although post-dating the original ICESCR, fall squarely within the commitments implied by this covenant.

The pragmatic recognition that resource limitations allow only 'progressive realization' of human rights renders it difficult or impossible for individuals, communities, or their advocates to achieve substantive change through the courts. The limited degree of penetration of international declarations such as the human right to water into national constitutions and domestic legislation also limits their impact on the lives of citizens (Ssenyonjo, 2017).

The international adoption of the human rights to water and sanitation in 2010 (jointly) and in 2015 (separately) marked important milestones in campaigners' attempts to put WASH on the global map. Human rights arguments may be important, but realization of those rights can be painfully slow, as the former Special Rapporteur on the right to food, Hilal Elver, concluded in her outgoing report: 'despite the Sustainable Development Goal of "zero hunger" and malnutrition by 2030, the realization of the right to food remains a distant, if not impossible, reality for far too many' (Elver, 2020).

Politicians, leaders of nations, and policy-makers arrive at decisions not solely on the grounds of rational arguments or scientific evidence, both of which have inherent limitations, but also through 'values, experience and political necessity' (Goldman and Pabari, 2020: 13). Part of my reason for alluding to imagination in the title of this final chapter is that those values and experiences (and possibly political necessities, too) nourish the imaginations of politicians and leaders, just as much as they feed the imaginations of poets, artists, writers, and ordinary citizens. The Iranian-American author and professor Azar Nafisi (2005) has put this well: 'Only curiosity about the fate of others, the ability to put ourselves in their shoes, and the will to enter their world through the magic of imagination, creates this shock of recognition. Without this empathy there can be no genuine dialogue, and we as individuals and nations will remain isolated and alien, segregated and fragmented.' Although it may be that the arts and literature seem to be the most obvious domains of the imagination, putting oneself in another's shoes, empathy, and compassion are universal.

Beyond rational arguments centred on health, poverty, human rights, and accountability – all of which have their place – perhaps direct appeals to the imaginations, values, empathy, and emotions of politicians and leaders would add to their effectiveness. After all, it is well established that citizens tend to invest in their own sanitation and hygiene, not primarily for scientifically evidenced health and economic reasons, but for reasons far more based on emotions such as disgust, shame, pride, and the natural urge to care for the family. Politicians are people too.

To summarize, investment in rural water should make sense to national governments, donors, and taxpayers because of (a) strong rational arguments around health and economy, (b) the shared global recognition of the rightness and justice of the case, and (c) shame, pride, and empathy among national and local leaders. Perhaps it is the third of these factors that adds a dimension of humanity to the rationality of the other arguments.

Getting the vision, mission, and values right

It is easy to be cynical about the value of organizational statements of vision (aspirations or ambitions), mission (the strategy used to get there), and values (the code of ethics, what the organization believes in). At the time of the International Decade of Water in the 1980s, such statements were relatively unusual; as we progress through the second such 'decade' – the International Decade for Action on Water for Sustainable Development (2018–28) towards the close of the SDG era – they are the norm. It could be the case that such statements have more to do with public relations – in the worst sense – than the realities of how organizations actually live, breathe, and operate. And yet, in the then ground-breaking management research *In Search of Excellence*, the authors stated:

> Let us suppose that we were asked for one all-purpose bit of advice for management, one truth that we were able to distill from the excellent companies research. We might be tempted to reply, "Figure out your value system. Decide what your company *stands for*. What does your enterprise do that gives everyone the most pride? Put yourself out ten or twenty years in the future: what would you look back on with greatest satisfaction?" (Peters and Waterman, 1982: 279)

My book is written for those organizations and individuals who are working toward a common vision, using a variety of strategies, and driven by sets of values or beliefs, which all matter greatly. Writing on this same topic, Watson was forthright:

> This then is my thesis: I firmly believe that any organization, in order to survive and achieve success, must have a sound set of beliefs on which it premises all its policies and actions. Next, I believe that the most important single factor in corporate success is faithful adherence to those beliefs ... the basic philosophy, spirit, and drive of an organization have far more to do with its relative achievements than do technological or economic resources, organizational structure, innovation and timing ... (Watson, 1963, quoted in Peters and Waterman, 1982: 280)

If the vision in rural water supply is succinctly summed up in the international NGO Water for People's mantra 'everyone forever', what are the minimum core values and operational principles required of organizations and individuals working to achieve those equity and sustainability outcomes? I suggest that the following shortlist sets out the crucial components of a code of ethics or values statement for those working in rural water.

First, **all people are of equal value**, regardless of skin colour, religious or political beliefs, ethnic origin, socio-economic status, age, infirmity or illness, sexual identity and orientation, or any other identifiable difference. This is easy to state, but much harder to observe in practice in a world which discriminates on all these and many other grounds. However, it is fundamental to the topic of this book. It requires that organizations actively determine how and why certain groups or individuals are marginalized or, in the words of the SDG declaration, 'left behind'. It requires the application of approaches that can alleviate rather than reinforce difference and disadvantage.

Second, **sustained service provision is paramount**. Water supply improvements that fail to provide lasting services are of little value. Infrastructure and management arrangements that function for a while but are soon either abandoned by their users or fall into long-term or permanent disrepair are unacceptable, and possibly worse than no service at all. Fulfilling this value is also hard. Providing new handpumps, solar systems, and taps is more visible and exciting than keeping the water flowing – and yet it is the latter that really counts.

Third, **national government leadership and systems are those ultimately mandated to serve their people**. Foreign donors and implementing organizations should be compliant with national policies and systems, and not by-pass or disregard them. To the extent that national systems are weak, under-resourced, or otherwise thought to be deficient, this is a matter for external organizations to address in a collaborative and constructive manner together with like-minded partners. It is only in the cases in which governments are unable to fulfill their mandates, or have effectively broken down, that humanitarian intervention – nevertheless still with the assent of government – is justified. Once more, this value or principle can be hard to observe, especially for external organizations impatient to make a difference. However, real change takes time, and such impatience must be resisted.

Fourth, **organizations and individuals working in rural water should be committed to the highest professional standards**. Ways of working, certification of competence, and accountability for the quality of work done, must all pursue excellence – mediocrity or amateurism are not good enough. This principle should begin with individuals; there should be no excuse for any individual working in the sector failing to have (or be working toward) recognition by an appropriate professional institution that both certifies the individual and requires their continuing professional development.

Fifth, **individuals and organizations must be committed to understanding the outcomes of their efforts, learning from this self-evaluation and reflection, and adapting accordingly**. Probably one of the most important reasons for the failure of the sector to progress as rapidly as many would like is inertia – sticking with time-honoured ways of working, despite the fact that they are no longer (if they were ever) fit for purpose amid new contexts and challenges. In the working lives of most professional individuals, the balance between 'learning the ropes' and actually 'sailing the ship' alters over the life course. In the first decade or two the emphasis is on learning; the balance gradually shifts, but reflection and learning must never stop.

The list of core values, both generic and specific to rural water, could be expanded. However, it is my belief that these five, were they to be applied consistently and thoughtfully, would make a great difference to progress in the sector.

I have placed values centrally in the articulation of organizations' vision, mission, and values for three reasons. First, if genuinely shared by the individuals in an organization, they express its 'heart' better than more tangible but soul-less aspects such as organigrams and strategy documents. Second, I believe most if not all organizations working in rural water can share the common vision of 'everyone forever' and the way this is articulated in the goal of SDG 6.1 to 'achieve universal and equitable access to safe and affordable drinking water for all' by 2030. And third, the missions or strategies of different organizations must reflect their individuality at the level of detail, even if there is much value in shared over-arching strategic goals. The WASH Agenda for Change (a set of principles adopted, at the time of writing, by 14 INGOs) nicely balances this mix. The Agenda for Change is an alliance of agencies that sign up to the district-wide, systems strengthening approach that tackles policy, financing, and institutions. It promotes harmonized district level work. It works to strengthen national level systems in order to enable all districts in the countries to reach everyone and ensure that services continue forever.

What should national rural water supply efforts aim to achieve by 2030?

In determining the specific goals of individual nations in this final decade of the SDGs, ambition needs to be balanced with realism. It is clear – despite some organizations' protestations to the contrary – that 'safely managed' services (on-premises, available when needed, fulfilling national water quality standards) will not be enjoyed by all in the year 2030. Too many countries are too far off track and investing too little to make such an outcome possible. What, then, is possible?

Each country has its own mix of the proportions of the rural population served by surface water, unimproved groundwater sources, limited (but improved) services, basic service (improved and within 30 minutes round trip), and safely managed services. I believe that the following phased approach could usefully guide those nations in which more progress is needed to advance households and communities up the ladder of better services (especially the 49 least well served countries identified in Chapter 11).

Phase 1: Improved water sources for all, with effective arrangements for sustainable services

The first principle is that every household in every community that is reliant on unimproved water sources (surface water and unimproved groundwater) should be served with improved (engineered for protection, even if untreated) sources. Achieving this may be through the type of community water supply programmes assumed in this book, as well as through supported self-supply programmes (Sutton and Butterworth, 2021). The use of unimproved water sources is the water supply equivalent of open defecation (OD) in the sanitation sector. Indeed, consumption of water from unimproved sources is tantamount to the ingestion of human faeces. Like OD, unimproved water sources should be eliminated with urgency.

Simultaneously, priority should be given to ensuring that all the necessary quality assurance, monitoring, management, and financing arrangements are in place so that every improved water source and system continues to provide a service that lasts. If investment in this essential dimension of rural water supply is neglected, progress will inevitably stagnate and founder.

In some countries, limited resources may mean that little more can be achieved by 2030 than the two measures just outlined. Some nations may even struggle to invest sufficient funds to achieve these joint goals of **improved water sources for all, with effective arrangements for sustainable services**. Without substantial international cooperation, the rural populations of such countries will contribute to the statistics of those still left behind in the year 2030.

Phase 2: At least basic services for all with effective arrangements for sustainable services

Where it is possible to invest more extensively, the next priority should be to bring all households and communities to a level of service that is at least basic. This should start with those whose services are limited, moving them into the basic category by constructing more water points and systems (to shorten collection times); only when a high proportion of the population enjoys basic service, should significant investments be made to move them up to on-premises and fully safely managed services.

At the same time, as access statistics improve and as communities move up the service ladder, even more investment (both in real terms and proportionately) will need to be made in the sustainability of services. Without this, hard-won progress will be wasted.

Phase 3: From basic to on-premises and safely managed services

Once a high proportion of the population (I suggest at least 70 per cent, while acknowledging that this figure is somewhat arbitrary) have at least basic services, the focus of investment can, with some fairness, shift to on-premises supply. To accomplish the full requirements of safely managed services, considerably increased investment in monitoring of performance and water quality, management, and recurrent financing to keep these higher levels of service working will be required.

I illustrate this phased approach in Figure 12.1. This figure has been created using the JMP data on rural water service levels for one country, Uganda. The phase zero baseline shows the mix in the year 2000, but over a 17-year period, the use of unimproved (surface water and groundwater) sources halved, while the use of limited and basic services (both constituting improved sources) increased considerably, as can be seen in Table 12.1.

I would argue that the national priority for the remainder of the SDG period for Uganda should be as shown in phase 2, namely to eliminate the use of surface water and unimproved groundwater, while focusing on basic services and continuing limited progress with on-premises water. This would provide the basis for a subsequent phase 3 in which on-premises and safely managed services are progressively realized. As the nation moves through the phases, investment in monitoring, management, and recurrent costs will need to rise (right-hand y-axis), not necessarily in a linear fashion, but many-fold.

This phased approach emphasizes the dual priorities of getting households at the lower service levels on to and climbing the rural water ladder, while simultaneously investing ever-increasing resources into keeping services working. The details have to be informed by the particular mix of existing service access statistics, but the principles are clear. The question then is what needs to be in place nationally to enable countries that currently are at the phase 0 or phase 1 stage to move progressively to the right of Figure 12.1?

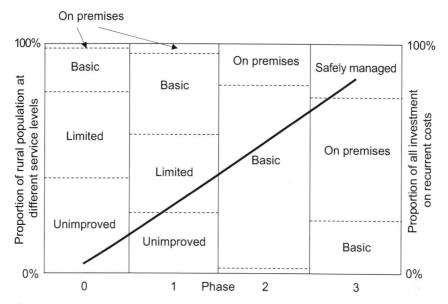

Figure 12.1 Schematic representation of a phased approach to rural water services, Uganda
Note: Phases 0 and 1 show actual situation in Uganda in 2000 and 2017 (Table 12.1); phases 2 and 3 show desirable trajectory over coming decades. Rising line (right hand y-axis) shows investments for sustainability

Table 12.1 Rural water service level access, 2000 and 2017, Uganda

Service level	2000	2017
Surface water	16.1%	8.4%
Unimproved groundwater	28.9%	14.5%
Limited service	35.8%	35.9%
Basic service	18.1%	36.8%
On premises	1.1%	4.5%

How can progress be achieved?

I suggest that the following five fundamental conditions need to be fulfilled in order for real and lasting progress to be made.

First, **national water policies** should explicitly give priority to eliminating open water use (surface water and unimproved/unprotected groundwater) and progressively moving communities to basic and on-premises water supply. National strategic plans should identify where in Figure 12.1 the nation's rural people currently sit, and set realistic targets for progression up the service ladder. National water policies should:

- focus on extending access, and moving communities up the service level ladder; while simultaneously ensuring that all existing and new

services are truly sustainable, and that all households and individuals are included;
- recognize and encourage a plurality of supply (self-supply and community systems) and management arrangements (household-managed, community-based, rural utility-managed);
- be absolutely clear about the division of responsibilities of water users and the state to finance the capital and recurrent costs of rural water supply;
- include in-built monitoring, learning and adaptation processes, so that services continue to function even when circumstances change.

Second, **national budget allocations** should rise accordingly, while focusing on predictability of such rises (rather than lurching unpredictably from one year's allocation to the next), and explicitly anticipating increased allocations, both proportionately and in real terms, for sustainability provisions – monitoring, management, and recurrent financial support. Budgets should clearly distinguish between the various cost components (especially capex, opex, and capmanex – see Chapter 8), and most crucially be explicit about the cost sharing of each component between water users and the public purse.

Third, **decentralization of governance** from national control to local governments should proceed fully. In many countries a degree of responsibility is handed down to local governments, but without the necessary financial resources, personnel, and full autonomy to fulfill their mandate for local services. Local governments will never be fully effective without such investment of people and resources, and it will take time for previously ineffectual institutions to grow and develop into their full capacities.

Fourth, **compliance and coordination** of all players, both at local level and nationally, are essential. External funders and implementing agencies must comply with government policies and systems, and at local level all the various players should follow a single plan that is constantly under review.

Fifth, **local governments as the de facto authorities, regulators, and monitors** of rural water services should hold up-to-date asset registers, implement routine monitoring of the services received by water users, undertake asset replacement according to pre-determined schedules and the findings of monitoring arrangements, regulate service providers and tariffs, and have some freedom to introduce local modifications to national policies.

Finally, all those involved in planning, financing and implementing rural water services should be **continuously learning and developing their knowledge and skills**. When all else is said, it is the individuals who work in the water sector who are key to its success.

External actors – donors and implementing organizations – must support such a vision, or risk perpetuating the problem rather than contributing to its solution.

Systems, complexity, and wickedness

At the end of Chapter 11 I referred to the relatively recent applications of complexity science and systems thinking to WASH, and specifically to rural water. A number of interrelated ideas have emerged during the 2000s and 2010s, some of which are more useful than others.

The first idea is that the arrangements for getting sustainable water services to rural communities involve numerous actors (national and local government, external donors, implementing non-governmental organizations (NGOs), local NGOs, private sector, and civil society), aspects (social and cultural, institutional, economic and financial, technical, and environmental), and processes (planning, budgeting, participation, monitoring, and communication). Together, these actors, aspects, and processes combine and interact to deliver services (or fail to do so). Together, this is the *system*.

The second idea is that the system is not merely *complicated*, but it is *complex*, in the sense outlined in Box 11.2. An important implication of such a notion is that predetermined plans will most probably not lead to the intended outcomes, due to the unpredictable nature of complex systems. We can 'prod' the system, but how it will respond is indeterminate.

The third idea is that of the 'wicked' problem. In the original paper the authors put the issue in this way:

> By now we are all beginning to realize that one of the most intractable problems is the problem of defining problems (of knowing what distinguishes an observed condition from a desired condition) and of locating problems (finding where in the complex causal networks the trouble really lies). In turn, and equally intractable, is the problem of identifying the actions that might effectively narrow the gap between what-is and what-ought-to-be. As we seek to improve the effectiveness of actions in pursuit of valued outcomes, as system boundaries get stretched, and as we become more sophisticated about the complex workings of open societal systems, it is becoming ever more difficult to make the planning idea operational. (Rittel and Webber, 1973: 159)

They went on to introduce the new term – wicked problem – thus: 'The kinds of problems that planners deal with – societal problems – are inherently different from the problems that scientists and perhaps some classes of engineers deal with. Planning problems are inherently wicked' (Rittel and Webber, 1973:160). Since this idea was first introduced, it has spawned an immense body of literature, some in support of the concept and some offering robust critique.

How useful are these ideas?

I believe it necessary for those working in the rural water professions to be clear about these three ideas and their applicability in our work.

First, is it useful to talk about the combination of sector actors, the issues that they have to address, and the processes that are undertaken as a *system*? In my view, the answer is a clear yes. There are many identifiable parts, there are clearly interactions and feedback loops between them, and it is an undoubted mental challenge to appreciate and understand the intricacy of the entire whole. I have argued elsewhere (Carter, 2016) that such understanding may not be fully acquired through scientific or social-scientific analysis but only through the tacit knowledge that comes with living with the system over a significant period of time – or as Meadows (2008) urged her readers, 'get the beat of the system'.

Second, is the WASH system, or the rural water system in particular, correctly described as a *complex* system? I would argue that this question is far less straightforward to answer. To give an analogy, I would have little hesitation in describing the human body and its health as a complex system. However, the unpredictability of its responses are far from total. If I break my arm, I can be fairly confident that a competent orthopaedic surgeon will be able to re-set the bones, fix them in place with plaster, and that, given sufficient time, healing will take place. A problem near to the simple end of the simple–complicated–complex spectrum can be readily solved. Of course there are far more intractable health problems, which neither Western nor any other medical systems can solve. Medical science proceeds by pursuing increasingly complete understanding, so pushing more and more complex health problems into the 'merely' complicated category.

I would therefore urge water professionals to question the idea of complexity as applied to their work. Some problems can be solved. For example, quality assurance systems for replacement components of physical water supply infrastructure have worked in the past, and they could do so once more. Other problems are more difficult. Community management of rural water supply is indeed challenging, but with the right support, it can be made to perform reasonably well. Even the difficult problems are not necessarily truly complex. The issue of full and reliable financing of rural water services seems to be insuperable at present, but even this may change.

Finally, what about 'wicked' problems? Despite the appeal for some of the phrase and the idea, even in the WASH sectors (see, for example, Casella et al., 2015), the pessimism of the original proponents of the idea is potentially paralyzing. Rittel and Webber (1973) conclude their paper with a cry of despair that such problems may simply be insoluble. The authors helped to expose the intractable nature of many social problems in a world of increasing plurality of beliefs and increasing challenge to the monopoly on expertise claimed by 'experts' in the past. However, although our task is indeed difficult, and in some respects complex, often it is merely complicated. It undoubtedly requires professional excellence; but without a measure of optimism, no progress will be made at all.

Systems strengthening

A number of international organizations are now committed to identifying the weaknesses in national WASH systems, and working with governments and others to strengthen those systems. WaterAid's SusWASH programme (Casey and Crichton-Smith, 2020), IRC's (2019) systems strengthening approach, and UNICEF's (2016) strengthening enabling environment are three instances of such approaches. A concise outline of the approach is that given by WaterAid:

> WASH system strengthening is about strengthening the environment into which WASH services and behaviours are introduced to ensure they continue to deliver benefits to everyone in society long after implementation. WASH system strengthening has evolved from an understanding that service delivery alone, without support to strengthen citizens' voice, government leadership and accountability, and public institutions, will fail to reach everyone in society with sustainable, high-quality WASH access. System strengthening requires detailed context analyses to identify areas of the system requiring targeted support. It takes time, requiring close collaboration with government (at multiple levels), service providers, civil society and communities. (SDG Partnership Platform, n.d., a)

All the approaches used by different organizations attempt in varying ways to address obstacles to progress jointly identified by government and external partners. All approaches have to adopt pragmatic compromises between theoretically rigorous systems thinking and practically useful ways of working. While acknowledging the interlinkages between the component parts – a key aspect of systems thinking – in practice they tend to work with the individual components or building blocks in a reductionist manner (Huston and Moriarty, 2018b; Valcourt, Javernick-Will et al., 2020). For example, IRC's nine building blocks (Huston and Moriarty, 2018b: 18) are all individually key components of the WASH system. However, it is their inter-linkages – such as how policy and legislation affect planning and budgeting; how finance affects monitoring; or how water supply infrastructure and its use impacts water resources – that turn a set of building blocks into a true system.

Two points with which to conclude this section: first, I believe much remains to be done to fully incorporate systems thinking into practical approaches for accelerating progress. Present approaches tend to the reductionistic, while the alternatives (such as comprehensive causal loop diagrams and the like) are perceived to be impractical.

Second, I believe the ideal should be for national institutions, with little or no external intervention, to undertake their own analysis of barriers to progress and potential solutions. Foreign aid for service delivery may already be decreasing; aid for institutional thinking may go the same way; even if this were not so, arguments around national autonomy and leadership would make it imperative for nations' dependence on external agencies to reduce.

Concluding remarks

Imagine a world in which people, especially young girls and women, no longer have to risk physical injury through hauling heavy loads of water, nor risk physical or sexual assault while fulfilling their water-fetching duties. Imagine a world in which households and whole communities are freed from the burdens imposed by uncertainties about the safety and reliability of water supply – freed to develop and flourish. Imagine whole districts of nations, free not only of open defecation but also of the ingestion of faecal pathogens caused by dependence on unprotected surface and groundwater sources. Districts that should rightly be proud of their state of development.

Imagine even the poorest nations investing as much as they can, as intelligently as they can, to these ends. Imagine external donors and organizations playing their parts, alongside national governments and others, to be part of the solution and never part of the problem. Imagine households, communities, local and national leaders, external agencies, and all those working for sustained and equitable, safe, reliable, and affordable rural water supply, all driven by the same can-do mentality that put two human beings on the moon, and returned them safely to earth, more than 40 years before the human right to safe and sustainable drinking water was declared.

In a 2018 interview, the influential thinker Robert Chambers (2018) urged those working for 'good change' to embrace love and empathy as their guiding principles. He approvingly quoted Martin Luther King Jr. – speaking in Atlanta, Georgia, in 1967 on the same theme – that powerful love is ultimately the only answer to mankind's problems, implementing the demands of justice (King, 1967). Could it be that a refreshed imagination, a renewed optimism that progress can be made, and an explicit focus on humanitarian values alongside data and evidence, could result in the better world – amply provided with sustainable water services for all – that we all wish to see?

This dream can be brought into reality. Many truisms have to be repeated: greater political will is needed; more investment is required; the best intelligence and professionalism must be brought to bear; all players must collaborate effectively. But together we truly can create a world that we prefer to live in. The challenge is difficult; it may have attributes of complexity; but without a healthy dose of optimism, nothing can be achieved.

ENDNOTE
National WASH systems sit within a global system of injustice

Like many other books addressing development challenges in low- and middle-income countries, this book acknowledges the realities of global inequalities, without offering any significant exploration of the reasons why some countries are very wealthy and others remain persistently mired in poverty.

Among the underlying assumptions of this book are the proposition that slow progress in the rural water sector is due to the meagre assets of households and communities and the deficient resources and capacities of governments, which are compounded by insufficient national political will to undergo systemic change. In other words, the problems lie within, and not external to, poor nations.

And yet, there is ample evidence that poor nations are poor and remain poor because of an extensive set of historical and present-day injustices. Historically, deep injustices resulted from the excesses of colonialism, through which transfers of vast natural resource wealth were made to the former colonial powers, which are now among the wealthiest nations on the planet. Slavery and other forms of violence were routinely used to subdue the colonized nations.

Present-day injustices include the disadvantageous terms of trade that low-income countries suffer, the debt burden that those same nations face, and the undemocratic and opaque nature of some of the most powerful global institutions such as the World Bank and the International Monetary Fund. Such injustices render it almost impossible for poor nations to climb out of their poverty.

In short, although the water, sanitation, and hygiene (WASH) system at country level indeed needs strengthening and reforming, failure to address deep injustices in the global economic system risks rendering those WASH systems strengthening efforts ineffective.

Unless and until there is a sufficient groundswell of public and political opinion in the wealthy and powerful nations to identify and address the true underlying causes of the poverty of nations and global inequality, little will change in national sectors such as rural water.

ANNEX
Some notes on definitions and statistics

Urban and rural

The definitions of 'urban' and 'rural' are not straightforward. In most publicly available statistics (for example UN data) that make the distinction between urban and rural populations, the definitions made by individual countries are used. The authors of the 2018 World Urbanization Prospects report carried out an analysis of the criteria used, summarizing their findings as follows:

> One hundred and twenty-one of the 233 countries or areas considered use administrative criteria to distinguish between urban and rural areas. Among these, 59 countries use administrative designations as the sole criterion. In 108 cases, the criteria used to characterize urban areas include population size or population density, and in 37 cases such demographic characteristics are the sole criterion. However, the lower limit above which a settlement is considered to be urban varies considerably, ranging between 200 and 50,000 inhabitants. Economic characteristics were part of the criteria used to identify urban areas in 38 countries or areas. Criteria related to functional characteristics of urban areas, such as the existence of paved streets, water-supply systems, sewerage systems or electric lighting, were part of the definition of urban in 69 cases, but only in eight cases were such criteria used alone. Lastly, in 12 cases there was no definition or an unclear definition of what constitutes the urban environment and in 12 cases the entire population of a country or area was considered to be urban. (UN DESA, 2019c)

The use of a range of country-based definitions leads to a significant degree of non-comparability between countries. As a consequence, there have been recent attempts to promote a single global set of definitions of urban and rural in order to permit global like-for-like comparisons. In March 2020 the UN Statistical Commission endorsed the Degree of Urbanization as a recommended method for international comparisons (EC et al., 2020). This method adopts simple thresholds of size and density of population, applied to 1 km² grid cells defined by geographic information system coding. The method is rendered workable by the availability of satellite-acquired data and analyses.

The Degree of Urbanization approach identifies three types of settlements:

- Cities, which have a population of at least 50,000 inhabitants in contiguous dense grid cells (>1,500 inhabitants per km²).

- Towns and semi-dense areas, which have a population of at least 5,000 inhabitants in contiguous grid cells with a density of at least 300 inhabitants per km^2.
- Rural areas, which consist mostly of low-density grid cells.

No doubt standardized and globally comparable definitions of urban and rural settings will be used increasingly in future; however, as most available data uses the existing country-specific definitions, I continue to use them in this book.

Country categories and terminology

This book focuses on countries that, in the past, might have been referred to as 'developing', 'less developed', or even 'third world'. All such terminology carries historical baggage or value judgments, and for that reason I try to avoid it here.

International organizations such as the United Nations and the World Bank use a number of different categorizations of countries (see, for example, UN DESA, 2019a). The World Bank deliberately moved away from the 'developing/developed' country dichotomy some years ago, and it now mainly uses a simple four-way classification by national income (GNI per capita, calculated by the World Bank Atlas method). The four categories are (a) low-income (<$1,025), (b) lower-middle income ($1,026–$3,995), (c) upper-middle income ($3,996–$12,375), and high-income (>$12,375). The 2020 classification of low-income and lower-middle income countries is shown here as Table A.1.

Table A.1 World Bank country classifications: low- and middle-income countries

Low income (n = 29)	Lower middle income (n = 50)	Lower middle income (cont.)
Afghanistan	Angola	Myanmar
Burkina Faso	Algeria	Nepal
Burundi	Bangladesh	Nicaragua
Central African Republic	Benin	Nigeria
Chad	Bhutan	Pakistan
Congo, Dem. Rep.	Bolivia	Papua New Guinea
Eritrea	Cabo Verde	Philippines
Ethiopia	Cambodia	São Tomé and Príncipe
Gambia, The	Cameroon	Senegal
Guinea	Comoros	Solomon Islands
Guinea-Bissau	Congo, Rep.	Sri Lanka
Haiti	Côte D'Ivoire	Tanzania
Korea, Dem. People's Rep.	Djibouti	Timor-Leste

Low income (n = 29)	Lower middle income (n = 50)	Lower middle income (cont.)
Liberia	Egypt, Arab Rep.	Tunisia
Madagascar	El Salvador	Ukraine
Malawi	Eswatini	Uzbekistan
Mali	Ghana	Vanuatu
Mozambique	Honduras	Vietnam
Niger	India	West Bank and Gaza
Rwanda	Kenya	Zambia
Sierra Leone	Kiribati	Zimbabwe
Somalia	Kyrgyz Republic	
South Sudan	Lao PDR	
Sudan	Lesotho	
Syrian Arab Republic	Mauritania	
Tajikistan	Micronesia, Fed. States	
Togo	Moldova	
Uganda	Mongolia	
Yemen, Rep.	Morocco	

Source: World Bank, 2020b; 2021 fiscal year

The UN system utilizes a variety of groupings, including:

- developed economies, economies in transition, and developing economies;
- groupings according to GNI per capita (the same as World Bank);
- least developed countries;
- heavily indebted poor countries;
- small island developing states;
- landlocked developing countries; and
- fuel exporting countries.

Both organizations also use geographical regions for some of their analyses and reporting, and these are not identical.

In the Sustainable Development Goals (SDG) era, the geographical classification of countries includes eight regions, further sub-divided into 21 sub-regions (Table A.2).

My interest in this book is in countries or sub-populations in countries where, so far, reliable and safe water supply services do not yet extend to the entire population. In principle this would include even some of the wealthiest countries such as Canada, the United States, and Australia, where services to indigenous people can be far from adequate and where failure to pay water bills can lead to disconnections, with all the consequential suffering that involves.

Table A.2 The SDG geographical regions

Region	Sub-regions	Countries
Sub-Saharan Africa	Eastern Africa	Burundi, Comoros, Djibouti, Eritrea, Ethiopia, Kenya, Madagascar, Malawi, Mauritius, Mayotte, Mozambique, Réunion, Rwanda, Seychelles, Somalia, South Sudan, Uganda, Tanzania, Zambia, Zimbabwe
	Middle Africa	Angola, Cameroon, Central African Republic, Chad, Congo, Democratic Republic of the Congo, Equatorial Guinea, Gabon, São Tomé and Príncipe
	Southern Africa	Botswana, Eswatini, Lesotho, Namibia, South Africa
	Western Africa	Benin, Burkina Faso, Cabo Verde, Côte d'Ivoire, Gambia, Ghana, Guinea, Guinea-Bissau, Liberia, Mali, Mauritania, Niger, Nigeria, Saint Helena, Senegal, Sierra Leone, Togo
Northern Africa and Western Asia	Northern Africa	Algeria, Egypt, Libya, Morocco, Sudan, Tunisia, Western Sahara
	Western Asia	Armenia, Azerbaijan, Bahrain, Cyprus, Georgia, Iraq, Israel, Jordan, Kuwait, Lebanon, Oman, Qatar, Saudi Arabia, State of Palestine, Syrian Arab Republic, Turkey, United Arab Emirates, Yemen
Central and Southern Asia	Central Asia	Kazakhstan, Kyrgyzstan, Tajikistan, Turkmenistan, Uzbekistan
	Southern Asia	Afghanistan, Bangladesh, Bhutan, India, Iran, Maldives, Nepal, Pakistan, Sri Lanka
Eastern and South-eastern Asia	Eastern Asia	China, Hong Kong SAR (China), Macao SAR (China), Taiwan Province of China, Dem. People's Republic of Korea, Japan, Mongolia, Republic of Korea
	South-eastern Asia	Brunei Darussalam, Cambodia, Indonesia, Lao People's Democratic Republic, Malaysia, Myanmar, Philippines, Singapore, Thailand, Timor-Leste, Vietnam
Latin America and the Caribbean	Caribbean	Anguilla, Antigua and Barbuda, Aruba, Bahamas, Barbados, Bonaire, Sint Eustatius and Saba, British Virgin Islands, Cayman Islands, Cuba, Curaçao, Dominica, Dominican Republic, Grenada, Guadeloupe, Haiti, Jamaica, Martinique, Montserrat, Puerto Rico, Saint Barthélemy, Saint Kitts and Nevis, Saint Lucia, Saint Martin (French part), Saint Vincent and the Grenadines, Sint Maarten (Dutch part), Trinidad and Tobago, Turks and Caicos Islands, United States Virgin Islands
	Central America	Belize, Costa Rica, El Salvador, Guatemala, Honduras, Mexico, Nicaragua, Panama
	South America	Argentina, Bolivia, Brazil, Chile, Colombia, Ecuador, Falkland Islands (Malvinas), French Guiana, Guyana, Paraguay, Peru, Suriname, Uruguay, Venezuela
Australia/New Zealand		Australia, New Zealand

Region	Sub-regions	Countries
Oceania (excl. Aus/NZ)	Melanesia	Fiji, New Caledonia, Papua New Guinea, Solomon Islands, Vanuatu
	Micronesia	Guam, Kiribati, Marshall Islands, Micronesia (Fed. States of), Nauru, Northern Mariana Islands, Palau
	Polynesia	American Samoa, Cook Islands, French Polynesia, Niue, Samoa, Tokelau, Tonga, Tuvalu, Wallis and Futuna Islands
Europe and Northern America	Eastern Europe	Belarus, Bulgaria, Czech Republic, Hungary, Poland, Republic of Moldova, Romania, Russian Federation, Slovakia, Ukraine
	Northern Europe	Channel Islands, Denmark, Estonia, Faroe Islands, Finland, Iceland, Ireland, Isle of Man, Latvia, Lithuania, Norway, Sweden, United Kingdom
	Southern Europe	Albania, Andorra, Bosnia and Herzegovina, Croatia, Gibraltar, Greece, Holy See, Italy, Malta, Montenegro, North Macedonia, Portugal, San Marino, Serbia, Slovenia, Spain
	Western Europe	Austria, Belgium, France, Germany, Liechtenstein, Luxembourg, Monaco, Netherlands, Switzerland
	Northern America	Bermuda, Canada, Greenland, Saint Pierre and Miquelon, United States of America

Source: UN DESA, 2019b

However, my main interest is in those countries and regions where genuine limitations of financial resources and weaknesses in governance, institutions, and capacity mean that water supply access is incomplete; and even where people are 'served', the service they experience is unreliable, unsafe, or otherwise inadequate.

Consequently I refer to countries of interest and country categories by a variety of terms. Where I refer to low-income or middle-income countries, this denotes the World Bank classifications. Sometimes I refer to 'lower-income' countries, meaning (more or less strictly) the low-income and lower-middle income countries.

References

Abramovsky, L., Andrés, L., Joseph, G., Rud, J.P., Sember, G. and Thibert, M. (2020) 'Study of the distributional performance of piped water consumption subsidies in 10 developing countries', World Bank Policy Research Working paper 9245, Water Global Practice, May. <https://openknowledge.worldbank.org/bitstream/handle/10986/33757/Study-of-the-Distributional-Performance-of-Piped-Water-Consumption-Subsidies-in-10-Developing-Countries.pdf?sequence=1&isAllowed=y> [accessed 25 September 2020].

Action contre le faim (ACF) (2005) *Water, Sanitation and Hygiene for Populations at Risk*, Paris: Hermann. <https://www.ircwash.org/sites/default/files/acf-2005-water.pdf> [accessed 12 September 2020].

Adekile, D. (2014a) 'Procurement and contract management of drilled well construction: a guide for supervisors and project managers', RWSN Publication 2014-10, St Gallen, Switzerland. <https://www.rural-water-supply.net/_ressources/documents/default/1-431-34-1418981862.pdf> [accessed 12 September 2020].

Adekile, D. (2014b) 'Supervising of water well drilling: a guide for supervisors', RWSN Publication 2014-5, St Gallen, Switzerland. <https://www.rural-water-supply.net/_ressources/documents/default/1-392-34-1418981410.pdf> [accessed 12 September 2020].

Alegre, H., Bapista, J.M., Cabrera, E.Jr., Cubillo, F., Duarte, P., Hirner, W., Merkel, W. and Parena, R. (2016) *Performance Indicators for Water Supply Services*, 3rd edn, London: IWA Publishing <https://doi.org/10.2166/9781780406336>.

Alexander, K., Tesfaye, Y., Dreibelbis, R., Abaire, B. and Freeman, M.C. (2015) 'Governance and functionality of community water schemes in rural Ethiopia', *International Journal of Public Health* 60 (8): 977–86 <https://doi.org/10.1007/s00038-015-0675-x>.

Alkire, S., Kanagaratnam, U. and Suppa, N. (2020) 'The global multidimensional poverty index (MPI) 2020', MPI Methodological note 49, OPHI, University of Oxford, July. <https://www.ophi.org.uk/wp-content/uploads/OPHI_MPI_MN_49_2020.pdf> [accessed 28 September 2020].

American Society for Testing and Materials (ASTM) (2015) 'Standard guide for development of groundwater monitoring wells in granular aquifers', D5521/D5521M-13. <https://www.astm.org/Standards/D5521.htm> [accessed 12 September 2020].

Anderson, M., Woessner, W.W. and Hunt R.J. (2015) *Applied Groundwater Modelling: Simulation of Flow and Advective Transport*, London: Academic Press.

Anderson, M.B., Brown, D. and Jean, I. (2012) *Time to Listen: Hearing People on the Receiving End of International Aid*, Cambridge, MA: CDA Collaborative Learning Projects. <https://www.cdacollaborative.org/publication/time-to-listen-hearing-people-on-the-receiving-end-of-international-aid/> [accessed 14 July 2020].

Andres, L., Boateng, K., Borja-Vega, C. and Thomas, E. (2018) 'A review of in-situ and remote sensing technologies to monitor water and sanitation interventions', *Water* 10 (6): 756 <https://doi.org/10.3390/w10060756>.

Andrews, M., Pritchett, L. and Woolcock, M. (2012) 'Escaping capability traps through problem-driven iterative adaptation (PDIA)', Center for Global Development Working paper 299. <http://www.cgdev.org/content/publications/detail/1426292> [accessed 16 July 2020].

Anríquez, G. and Stloukal, L. (2008) 'Rural population change in developing countries: lessons for policymaking', *European View* 7: 309–17 <https://doi.org/10.1007/s12290-008-0045-7>.

Appelo, T. (2006) 'Arsenic in groundwater: a world problem', Proceedings Seminar, Utrecht, 29 November. <http://documentacion.ideam.gov.co/openbiblio/bvirtual/021448/Titulo.pdf> [accessed 14 September 2020].

Aqua Publica Europea (2016) 'Water affordability: public operators' views and approaches on tackling water poverty'. <https://www.aquapublica.eu/sites/default/files/document/file/ape_water_affordability_final_0.pdf> [accessed 24 September 2020].

Arlosoroff, S., Tschannerl, G., Grey, D., Journey, W., Karp, A., Langenegger, O. and Roche, R. (1987) *Community Water Supply: The Handpump Option*, Washington, DC: World Bank. <http://documents.worldbank.org/curated/en/299321468765272889/pdf/multi0page.pdf> [accessed 7 July 2020].

Armstrong, A., Mahan, J. and Zapor, J. (2017) 'Solar pumping for rural water supply: life-cycle costs from eight countries', Paper presented at the 40th WEDC International Conference, Loughborough, 24–28 July. <https://wedc-knowledge.lboro.ac.uk/resources/conference/40/Armstrong-2654.pdf> [accessed 18 September 2020].

Arnalich, S. (2009) 'How to design a gravity flow water system through worked exercises' [online], 1 June. <https://issuu.com/arnalich/docs/ligraxen> [accessed 18 September 2020].

Arnalich, S. (2010) 'Gravity flow water supply: conception, design and sizing for cooperation projects' [online], 1 June. <https://issuu.com/arnalich/docs/ligraven> [accessed 18 September 2020].

Arnalich, S. (2011a) 'Epanet and development: a progressive 44 exercise workbook' [online]. <https://gumroad.com/l/uIbex> [accessed 18 September 2020].

Arnalich, S. (2011b) 'Epanet and development: how to calculate water networks by computer' [online]. <https://gumroad.com/l/oaSfD> [accessed 18 September 2020].

Arthur, R. (2019) 'UK bottled water sales reach 4bn litres, with "robust growth" to continue', *Beverage Daily*. <https://www.beveragedaily.com/Article/2019/03/14/UK-bottled-water-sales-reach-4bn-litres-with-robust-growth-to-continue> (published 14 March) [accessed 7 July 2020].

Aunger, R., Greenland, K., Ploubidis, G., Schmidt, W., Oxford, J. and Curtis, V. (2016) 'The determinants of reported personal and household hygiene behaviour: a multi-country study', *PloS One* 11 (8): e0159551 <https://doi.org/10.1371/journal.pone.0159551>.

Bacon F. (1601) *The Essays or Counsels, Civil and Moral, of Francis Lord Verulam, Viscount St. Albans*. <https://www.fulltextarchive.com/pdfs/Essays.pdf> [accessed 26 September 2020].

Bader, H-P., Berg, M., Bretzler, A., Gebauer, H., Huber, A.C., Hug, S.J., Inauen, J., Johnson, C.A., Johnston, R.B., Lüthi, C., Khan N., Mosler, H-J., Osterwalder, L., Roberts, L.C., Scheidegger, R., Tobias, R. and Yang, H. (2017) *Geogenic Contamination Handbook - Addressing Arsenic and Fluoride in Drinking Water*, Dübendorf: Swiss Federal Institute of Aquatic Science and Technology (Eawag). <https://www.eawag.ch/fileadmin/Domain1/Forschung/Menschen/Trinkwasser/Wrq/Handbook/geogenic-contamination-handbook.pdf> [accessed 10 September 2020].

Bain R., Cronk R., Hossain R., Bonjour S., Onda K., Wright J., Yang H., Slaymaker T., Hunter P., Prüss-Ustün A., Bartram J. (2014) 'Global assessment of exposure to faecal contamination through drinking water based on a systematic review', *Tropical Medicine and International Health* 19 (8): 917–27 <https://doi.org/10.1111/tmi.12334>.

Banks, B. and Furey, S. (2016) 'What's working, where, and for how long: a 2016 water point update', Paper presented at the 7th RWSN Forum, Abidjan, Côte d'Ivoire, November 29 – 2 December. <https://www.rural-water-supply.net/_ressources/documents/default/1-787-2-1502962732.pdf> [accessed 17 September 2020].

Bartram, J., Brocklehurst, C., Fisher, M.B., Luyendijk, R., Hossain, R., Wardlaw, T. and Gordon, B. (2014) 'Global monitoring of water supply and sanitation: history, methods and future challenges', *International Journal of Environmental Research and Public Health* 11 (8): 8137–65 <https://doi.org/10.3390/ijerph110808137>.

Bartram, J., Corrales, L., Davison, A., Deere, D., Drury, D., Gordon, B., Howard, G., Rinehold, A. and Stevens, M. (2009) *Water Safety Plan Manual: Step-by-Step Risk Management for Drinking-Water Suppliers*, Geneva: WHO. <https://apps.who.int/iris/bitstream/handle/10665/75141/9789241562638_eng.pdf?sequence=1> [accessed 10 September 2020].

Bates I., Chabala L.M., Lark R.M., MacDonald A., Mapfumo P., Mtambanengwe F., Nalivata P.C., Owen R., Phiri E., Pulford J. (2020) 'Letter to the Editor: response to global soil science research collaboration in the 21st century: time to end helicopter research by Minasny et al.', *Geoderma* 378: 114559 <https://doi.org/10.1016/j.geoderma.2020.114559>.

Baumann, E. (2006) 'Do operation and maintenance pay?' *Waterlines* 25 (1): 10–12. <https://www.ircwash.org/sites/default/files/Baumann-2006-Do.pdf> [accessed 20 September 2020].

Baumann, E. and Furey, S.G. (2013) 'How three handpumps revolutionised rural water supplies: a brief history of the India Mark II/III, Afridev and the Zimbabwe Bush Pump', Field note No. 2013-1, Skat, RWSN, St Gallen, Switzerland. <https://www.rural-water-supply.net/_ressources/documents/default/1-475-2-1363951079.pdf> [accessed 17 September 2020].

Bianchi, M., MacDonald, A.M., Macdonald, D. and Asare E.B. (2020) 'Investigating the productivity and sustainability of weathered basement aquifers in tropical Africa using numerical simulation and global sensitivity analysis', *Water Resources Research* 56 (9): e2020WR027746 <https://doi.org/10.1029/2020WR027746>.

Bird, K., Hulme, D., Moore, K. and Shepherd, A. (2001) 'Chronic poverty and remote rural areas', Chronic Poverty Research Centre Working paper No. 13. <https://assets.publishing.service.gov.uk/media/57a08d46e5274a27b200174d/13Bird_et_al.pdf> [accessed 26 September 2020].

Blackman, M. (2006) *Noughts and Crosses*. London: Corgi.
Bogle, J.C. (2010) *Enough: True Measures of Money, Business, and Life*, Hoboken, NJ: Wiley.
Bonan, G. (2019) 'Soil moisture', in G. Bonan (ed.), *Climate Change and Terrestrial Ecosystem Monitoring*, pp. 115–33, Cambridge: Cambridge University Press <https://doi.org/10.1017/9781107339217.009>.
Bongaarts, J. (2009) 'Human population growth and the demographic transition', *Philosophical Transactions of the Royal Society B* 364 (1532): 2985–90 <https://doi.org/10.1098/rstb.2009.0137>.
Bonsor, H., MacDonald, A.M., Casey, V., Carter, R. and Wilson, P. (2018) 'The need for a standard approach to assessing the functionality of rural community water supplies', *Hydrogeology Journal* 26 (2): 367–70 <https://doi.org/10.1007/s10040-017-1711-0>.
Boudet, M., María, A., Petesch, P., Turk, C. and Thumala, A. (2013) *On Norms and Agency: Conversations about Gender Equality with Women and Men in 20 Countries*, Washington, DC: World Bank <https://doi.org/10.1596/978-0-8213-9862-3>.
Bouman, F.J.A. (1977) 'Indigenous savings and credit societies in the third world – a message', *Savings and Development* 1 (4): 181–219 <https://www.jstor.org/stable/25829637>.
Brammer, H. and Ravenscroft, P. (2009) 'Arsenic in groundwater: a threat to sustainable agriculture in South and South-east Asia', *Environment International* 35: 647–54 <https://doi.org/10.1016/j.envint.2008.10.004>.
Brassington, R. (2007) *Field Hydrogeology*, 3rd edn, Chichester: Wiley.
Brick, J. (2008) 'The political economy of customary village organizations in rural Afghanistan', Paper presented at the Annual Meeting of the Central Eurasian Studies Society, Washington, DC, 18–21 September. <http://www.bu.edu/aias/brick.pdf> [accessed 26 September 2020].
Brioschi, C.A. (2017) *Corruption: A Short History*, Washington, DC: Brookings Institution Press.
British Geological Survey (BGS) (no date) 'Fluoride in groundwater' [online]. <https://www.bgs.ac.uk/research/groundwater/health/fluoride.html> [accessed 11 September 2020].
Broadbent, A., Walker, D., Chalkidou, K., Sullivan, R. and Glassman, A. (2020) 'Lockdown is not egalitarian: the costs fall on the global poor', *The Lancet* 396: 21–2 <https://doi.org/10.1016/S0140-6736(20)31422-7>.
Budge, S., Hutchings, P., Parker, A., Tyrrel, S., Tulu, T., Gizaw, M. and Garbutt, C. (2019) 'Do domestic animals contribute to bacterial contamination of infant transmission pathways? Formative evidence from Ethiopia', *Journal of Water and Health* 17 (5): 655–69 <https://doi.org/10.2166/wh.2019.224>.
Burr, P. and Fonseca, C. (2013) 'Applying a life-cycle costs approach to water: costs and service levels in rural and small town areas in Andhra Pradesh (India), Burkina Faso, Ghana and Mozambique', IRC WASHCost Working paper 8. <https://www.ircwash.org/sites/default/files/20130208_8_wp_water_web_2.pdf> [accessed 23 September 2020].
Cairncross, S. and Feachem, R. (1993) *Environmental Health Engineering in the Tropics: An Introductory Text*, 2nd edn, Chichester: John Wiley & Sons.
Cairncross, S. and Valdmanis, V. (2006) 'Water supply, sanitation, and hygiene promotion' in D.T. Jamison et al. (eds) *Disease Control Priorities in*

Developing Countries, 2nd edn, Washington, DC: The International Bank for Reconstruction and Development / The World Bank; New York, NY: Oxford University Press. <https://www.ncbi.nlm.nih.gov/books/NBK11755/> [accessed 7 July 2020].

The Carpenters. (1970) 'We've only just begun' [song], written by R. Nichols and P. Williams, featured on *Close To You*, A&M Records.

Carrard, N., Foster, T. and Willetts, J. (2019) 'Groundwater as a source of drinking water in Southeast Asia and the Pacific: a multi-country review of current reliance and resource concerns', *Water* 11 (8): 1605 <https://doi.org/10.3390/w11081605>.

Carroll, D. (1962) 'Rainwater as a chemical agent of geologic processes – a review', US Geological Survey Water Supply, Paper 1535-G, US Government Printing Office, Washington, DC. <https://pubs.usgs.gov/wsp/1535g/report.pdf> [accessed 12 September 2020].

Carter, R.C. (2015) 'Long live the humble handpump', Paper presented at the International Association of Hydrogeologists 42nd Annual Congress, Rome, September 2015. <https://upgro.files.wordpress.com/2015/09/carter-id-n168-session-s1-1a.pdf> [accessed 16 September].

Carter, R.C. (2016) 'Water in development', in C.M. Ainger and R.A. Fenner (eds) *Sustainable Water: Delivering Sustainable Infrastructure*, pp. 197–215, London: ICE Publishing.

Carter, R.C. (2019a) 'Keeping community-managed handpump systems going: all systems go!', Paper presented at WASH systems symposium, The Hague, The Netherlands, 12–14 March. <https://www.ircwash.org/sites/default/files/084-201906carter-rc.pdf> [accessed 7 July 2020].

Carter, R.C. (2019b) 'Leave no-one behind in rural water services', *Waterlines* 38 (2): 69–70 <http://dx.doi.org/10.3362/1756-3488.2019.38-2ED>.

Carter, R.C. and Ross, I. (2016) 'Beyond "functionality" of handpump-supplied rural water services in developing countries', *Waterlines* 35 (1): 94–110 <https://doi.org/10.3362/1756-3488.2016.008>.

Carter, R.C. and Rwamwanja, R. (2006) 'Functional sustainability in community water and sanitation: a case study from south-west Uganda', Diocese of Kigezi Church of Uganda, Cranfield University Silsoe, and Tearfund, August. <https://reliefweb.int/sites/reliefweb.int/files/resources/C6DBD957F6EF817FC12572E500412CA8-tearfund-water-dec06.pdf> [accessed 20 September 2020].

Carter, R.C., Chilton, J., Danert, K. and Olschewski, A. (2014) 'Siting of drilled water wells: a guide for project managers', RWSN Publication 2014-11, St Gallen, Switzerland. <https://www.rural-water-supply.net/_ressources/documents/default/1-187-4-1521131535.pdf> [accessed 12 September 2020].

Carter, R.C., Harvey, E. and Casey, V. (2010) 'User financing of rural handpump water services', Paper presented at IRC Symposium, Pumps, Pipes and Promises, The Hague, The Netherlands, 16–18 November 2010. <https://www.ircwash.org/sites/default/files/Carter-2010-User.pdf> [accessed 7 July 2020].

Casella, D., van Tongeren, S. and Nikolic, I. (2015) 'Change in complex adaptive systems: a review of concepts, theory and approaches for tackling "wicked" problems in achieving sustainable rural water services', IRC Working paper, December. <https://www.ircwash.org/resources/whole-system-approach-change-review-concepts-theory-and-approaches-tackling-'wicked'> [accessed 4 September 2020].

Casey, V., Brown, L., Carpenter, J.D., Nekesa, J. and Etti, B. (2016) 'The role of handpump corrosion in the contamination and failure of rural water supplies', *Waterlines* 35 (1): 59–77 <http://dx.doi.org/10.3362/1756-3488.2016.006>.

Casey, V., Crichton-Smith, H. (2020) 'System strengthening for inclusive, lasting WASH that transforms people's lives: practical experiences from the SusWASH programme', H&M Foundation and Wateraid. <https://washmatters.wateraid.org/sites/g/files/jkxoof256/files/suswash-global-learning-report.pdf> [accessed 11 November 2020].

Castañeda, A., Doan, D., Newhouse, D., Nguyen, M.C., Uematsu, H., Azevedo, J.P. and World Bank Data for Goals Group. (2018) 'A new profile of the global poor', *World Development* 101: 250–67 <https://doi.org/10.1016/j.worlddev.2017.08.002>.

Cavill, S., Francis, N., Grant, M., Huggett, C., Leahy, C., Leong, L., Mercer, E., Myers, J., Singeling, M. and Rankin, T. (2020) 'A call to action: organizational, professional, and personal change for gender transformative WASH programming', *Waterlines* 39 (2&3): 219–37 <http://dx.doi.org/10.3362/1756-3488.20-00004>.

Center for Economic and Social Rights (CESR) (no date) 'What are economic, social and cultural rights?' [online]. <https://www.cesr.org/what-are-economic-social-and-cultural-rights> [accessed 18 September 2020].

Centers for Disease Control and Prevention (CDC) (2011) 'The safe water system: safe storage of drinking water'. <https://www.cdc.gov/safewater/pdf/safestorage_2011-c.pdf> [accessed 11 September 2020].

Chambers, R. (2018) 'Can we know better? Reflections for development' [podcast interview], *Between the Lines*, Episode 1, 2 October. <https://www.ids.ac.uk/download.php?file=wp-content/uploads/2018/10/Ep-01-Robert-Chambers-podcast-transcript-1.pdf> [accessed 7 September 2020].

Chowns, E. (2015) 'Water point sustainability and the unintended impacts of community management in Malawi', paper presented at the 38th WEDC International Conference, Loughborough University, UK, 27–31 July. <https://repository.lboro.ac.uk/articles/Water_point_sustainability_and_the_unintended_impacts_of_community_management_in_Malawi/9586958/1> [accessed 20 September 2020].

Churchill, A.A. (1987) *Rural Water Supply and Sanitation: Time for a Change*, Washington, DC: World Bank.

Clark, A., Jit, M., Warren-Gash, C., Guthrie, B., Wang, H.H.X., Mercer, S.W., Sanderson, C., McKee, M., Troeger, C., Ong, K.L., Checchi, F., Perel, P., Joseph, S., Gibbs, H.P., Banerjee, A., Eggo, R.M., with the Centre for the Mathematical Modelling of Infectious Diseases COVID-19 working group (2020) 'Global, regional, and national estimates of the population at increased risk of severe COVID-19 due to underlying health conditions in 2020: a modelling study', *Lancet Global Health* 8 (8): E1003–17 <https://doi.org/10.1016/S2214-109X(20)30264-3>.

Cleaver, F. and Whaley, L. (2018) 'Understanding process, power, and meaning in adaptive governance: a critical institutional reading', *Ecology and Society* 23 (2): 49 <https://doi.org/10.5751/ES-10212-230249>.

Cooper-Vince, C.E., Kakuhikire, B., Vorechovska, D., McDonough, A.Q., Perkins, J., Venkataramani, A.S., Mushavi, R.C., Baguma, C., Ashaba, S., Bangsberg, D.R. and Tsai, A.C. (2017) 'Household water insecurity, missed schooling, and the mediating role of caregiver depression in rural Uganda', *Global Mental Health* 4 (15) <https://doi.org/10.1017/gmh.2017.14>.

Crane, R.J., Jones, K.D.J. and Berkley, J.A. (2015) 'Environmental enteric dysfunction: an overview,' *Food and Nutrition Bulletin* 36 (1): 76–87 <https://journals.sagepub.com/doi/pdf/10.1177/15648265150361S113>.

Cumming, O., Benjamin, F. Arnold, B.F., Ban, R., Clasen, T., Mills, J.E., Freeman, M.C., Gordon, B., Guiteras, R., Howard, G., Hunter, P.R., Johnston, R.B., Pickering, A.J., Prendergast, A.J., Prüss-Ustün, A., Rosenboom, J.W., Spears, D., Sundberg, S., Wolf, J, Null, C., Luby, S.P., Humphrey, J.H. and Colford, J.M. (2019) 'The implications of three major new trials for the effect of water, sanitation and hygiene on childhood diarrhea and stunting: a consensus statement', *BMC Medicine* 17 (173) <https://doi.org/10.1186/s12916-019-1410-x>.

Curtis, V. (1986) *Women and the Transport of Water*, Rugby: Intermediate Technology Publications.

Curtis, V. and Cairncross, S. (2003) 'Effect of washing hands with soap on diarrhoea risk in the community: a systematic review', *The Lancet Infectious Diseases* 3, 275–81 <https://doi.org/10.1016/s1473-3099(03)00606-6>.

Cuthbert, M.O., Gleeson, T., Moosdorf, N., Befus, K.M., Schneider, A., Hartmann, J. and Lehner, B. (2019) 'Global pattern and dynamics of climate-groundwater interactions', *Nature Climate Change* 9: 137–41 <https://doi.org/10.1038/s41558-018-0386-4>.

Cuthbert, M.O., Taylor, R.G., Favreau, G., Todd, M.C., Shamsudduha, M., Villholth, K.G., MacDonald, A.M., Scanlon, B.R., Kotchoni, D.O.V., Vouillamoz, J-M., Lawson, F.M.A., Adjomayi, P.A., Kashaigili, J., Seddon, D., Sorensen, J.P.R., Ebrahim, G.Y., Owor, M., Nyenje, P.M., Nazoumou, Y., Goni, I., Ousmane, B.I., Sibanda, T., Ascott, M.J., Macdonald, D.M.J., Agyekum, W., Koussoubé, Y., Wanke, H., Kim, H., Wada, Y., Lo, M.H., Oki., T. and Kukuric, N. (2019) 'Observed controls on resilience of groundwater to climate variability in sub-Saharan Africa', *Nature* 572 (7768): 230–4 <https://doi.org/10.1038/s41586-019-1441-7>.

Danert, K. (2015) 'Manual Drilling Compendium', RWSN Publication 2015-2, St Gallen, Switzerland. <https://www.rural-water-supply.net/_ressources/documents/default/1-653-34-1442223588.pdf> [accessed 12 September 2020].

Danert, K. and Hutton, G. (2020) 'Shining the spotlight on household investments for water, sanitation and hygiene (WASH): let us talk about HI and the three 'T's', *Journal of Water, Sanitation and Hygiene for Development* 10 (1): 1–4 <https://doi.org/10.2166/washdev.2020.139>.

Danert, K., Adekile, D. and Kanuto, J.G. (2020) 'Striving for borehole drilling professionalism in Africa: a review of a 16-year initiative through the Rural Water Supply Network from 2004 to 2020', *Water* 12 (2): 3305 <https://doi.org/10.3390/w12123305>.

Danert, K., Armstrong, T., Adekile, D., Duffau, B., Ouedraogo, I. and Kwei, C. (2010) 'Code of practice for cost effective boreholes', RWSN, St Gallen, Switzerland. <https://www.rural-water-supply.net/en/resources/details/128> [accessed 26 September 2020].

Danert, K., Furey, S., Mechta, M. and Gupta, S.K. (2016) 'Effective joint sector reviews for water, sanitation and hygiene (WASH): a study and guidance', World

Bank Group/WSP/Skat Consulting, January. <https://skat.ch/wp-content/uploads/2017/01/1-757-3-1463486911.pdf> [accessed 16 July 2020].

Danert, K., Luutu, A., Carter, R.C. and Olschewski, A. (2014) 'Costing and pricing: a guide for water well drilling enterprises', RWSN Publication 2014-12, St Gallen, Switzerland. <https://www.rural-water-supply.net/_ressources/documents/efault/1-146-34-1418982599.pdf> [accessed 12 September 2020].

Davies, J., Robins, N.S. and Farr, J. (2013) 'Is the Precambrian basement aquifer in Malawi up to the job?', in J. Cobbing, S. Adams, I. Dennis and K. Riemann (eds), *Assessing and Managing Groundwater in Different Environments*, pp. 241–50, London: CRC Press.

Davis, D.M. (2012) 'Socio-economic rights', in M. Rosenfeld and A. Sajó (eds) *The Oxford Handbook of Comparative Constitutional Law*, pp. 1020–35, Oxford: Oxford University Press. <https://doi.org/10.1093/oxfordhb/9780199578610.013.0051>.

Davis, J. and Lambert, R. (2002) Engineering in Emergencies: A Practical Guide for Relief Workers, 2nd edn, London: ITDG/RedR.

Davis, J., Crow, B. and Miles, J. (2012) 'Measuring water collection times in Kenyan informal settlements', Proceedings of the Fifth International Conference on Information and Communication Technologies and Development, ACM, Atlanta, GA. <https://users.soe.ucsc.edu/~davis/papers/2012_ICTD_p114-davis.pdf> [accessed 10 September 2020].

de Albuquerque, C. (2014) *Realising the Human Rights to Water and Sanitation: A Handbook*, Geneva: OHCHR. <http://www.rural-water-supply.net/en/resources/details/654> [accessed 26 September 2020].

de Asís, M.G., O'Leary, D., Ljung, P. and Butterworth, J. (2009) *Improving Transparency, Integrity, and Accountability in Water Supply and Sanitation: Action, Learning, Experiences*, Washington, DC: The World Bank Institute and Transparency International. <https://www.oecd.org/env/outreach/44475062.pdf> [accessed 14 July 2020].

de Waal, D., Hirn, M. and Mason, N. (2011) 'Pathways to progress: status of water and sanitation in Africa', African Ministers Council on Water/The World Bank/Water and Sanitation Program. <http://www.wsp.org/wsp/content/pathways-progress-status-water-and-sanitation-africa> [accessed 16 July 2020].

de Wiest, R.J.M. (1965) *Geohydrology*, New York, NY: Wiley.

Department for International Development (DFID) (2019) 'Governance for growth, stability and inclusive development', DFID Position paper, March. <https://assets.publishing.service.gov.uk/government/uploads/system/uploads/attachment_data/file/786751/Governance-Position-Paper2a.pdf> [accessed 9 July 2020].

Descroix, L., Mahé, G., Lebel, T., Favreau, G., Galle, S., Gautier, E., Olivry, J-C., Albergel, J., Amogu, O., Cappelaere, B., Dessouassi, R., Diedhiou, A., Le Breton, E., Mamadou, I. and Sighomnou, D. (2009) 'Spatio-temporal variability of hydrological regimes around the boundaries between Sahelian and Sudanian areas of West Africa: a synthesis', *Journal of Hydrology* 375: 90–102 <https://doi.org/10.1016/j.jhydrol.2008.12.012>.

Development Tracker (2020) 'Water, sanitation and hygiene results programme to support scale-up efforts' [online] (updated 31 March 2020). <https://devtracker.dfid.gov.uk/projects/GB-1-203572> [accessed 19 August 2020].

Döll, P. and Fiedler K. (2008) 'Global scale modelling of groundwater recharge', *Hydrology and Earth Systems Sciences* 12 (3): 863–85 <https://doi.org/10.5194/hess-12-863-2008>.

Drinking Water Inspectorate (DWI) (no date) 'Private water supplies in England and Wales' [online], (modified 23 March 2020). <http://www.dwi.gov.uk/private-water-supply/> [accessed 20 September 2020].

Driscoll, F.G. (1986) *Groundwater and Wells*, 2nd edn, St Paul, MN: Johnson Screens.

D'Souza, R. (2018) What's Wrong with Rights? Social Movements, Law and Liberal Imaginations, London: Pluto Press.

EC, ILO, FAO, OECD, UN-Habitat and the World Bank (2020) 'A recommendation on the method to delineate cities, urban and rural areas for international statistical comparisons', UN Statistical Commission 51st Session, item 3(j) of the provisional agenda. <https://unstats.un.org/unsd/statcom/51st-session/documents/BG-Item3j-Recommendation-E.pdf> [accessed 19 May 2020].

Edossa, D.C., Babel, M.S, Gupta, A.D. and Awulachew, S.B. (2005) 'Indigenous systems of conflict resolution in Oromia, Ethiopia', in B. van Koppen, M. Giordano, and J. Butterworth (eds) *Community-Based Water Law and Water Resource Management Reform in Developing Countries*, pp. 146–57, Silverton: IMWI <https://doi.org/10.1079/9781845933265.0146>.

EED Advisory (2018) 'Evaluation of the sustainability of solar powered water supply systems in Kenya', Nairobi. <https://www.rural-water-supply.net/_ressources/documents/default/1-820-2-1543415708.pdf> [accessed 15 September 2020].

Elver, H. (2020) 'Critical perspective on food systems, food crises and the future of the right to food', Report of the Special Rapporteur on the right to food Human Rights Council Forty-third session 24 February–20 March, Agenda item 3, A/HRC/43/44. <https://undocs.org/en/A/HRC/43/44> [accessed 2 September 2020].

Estevan, H. and Schaefer, B. (2017) 'Life cycle costing: state of the art report', ICLEI – Local Governments for Sustainability, European Secretariat, March. <https://sppregions.eu/fileadmin/user_upload/Life_Cycle_Costing_SoA_Report.pdf> [accessed 23 September 2020].

European Union (2014) 'Directive 2014/24/EU of the European Parliament and of the Council'. <https://eur-lex.europa.eu/eli/dir/2014/24/2020-01-01> [accessed 23 September 2020].

Evans, A. (2018) 'Cities as catalysts of gendered social change? Reflections from Zambia', *Annals of the American Association of Geographers* 108 (4): 1096–114.

Evans, A. (2019) 'How cities erode gender inequality: a new theory and evidence from Cambodia', *Gender and Society* 33 (6): 961–84 <https://doi.org/10.1177/0891243219865510>.

Falkenmark, M. (1995) 'Land-water linkages: a synopsis', in Land and Water Integration in River Basin Management, Proceedings of an FAO informal workshop, 31 January – 2 February 1993. <http://www.fao.org/3/v5400e/v5400e00.htm> [accessed 12 September 2020].

Fallas, H.C., MacDonald, A.M., Casey, V., Kebede, S., Owor, M., Mwathunga, E., Calow, R., Cleaver, F., Cook, P., Fenner, R.A., Dessie, N., Yehualaeshet, T., Wolde, G., Okullo, J., Katusiime, F., Alupo, G., Berochan, G., Chavula, G.,

Banda, S., Mleta, P., Jumbo, S., Gwengweya, G., Okot, P., Abraham, T., Kefale, Z., Ward, J., Lapworth, D., Wilson, P., Whaley, L. and Ludi, E. (2018) 'UPGro Hidden Crisis Research consortium: project approach for defining and assessing rural water supply functionality and levels of performance', British Geological Survey open report, OR/18/060. <http://nora.nerc.ac.uk/id/eprint/523090/> [accessed 14 September 2020].

Feachem, R. (1978) Water, Health and Development: An Interdisciplinary Evaluation, London: Tri-med Books Ltd.

Fetter, C.W. (1994) *Applied Hydrogeology*, 3rd edn, New York, NY: MacMillan.

Fink, G., D'Acremont, V., Leslie, H.H. and Cohen, J. (2019) 'Antibiotic exposure among children younger than 5 years in low-income and middle-income countries: a cross-sectional study of nationally representative facility-based and household-based surveys', *The Lancet Infectious Diseases* 20 (2): 179–87 <https://doi.org/10.1016/S1473-3099(19)30572-9>.

Fisher, M.B., Shields, K.F., Chan, T.U., Christenson, E., Cronk, R., Leker, H., Samani, D., Apoya, P., Lutz, A. and Bartram, J. (2015) 'Understanding handpump sustainability: determinants of rural water source functionality in the Greater Afram Plains region of Ghana', *Water Resources Research* 51: 8431–49 <https://doi.org/10.1002/2014WR016770>.

Food and Agriculture Organization of the United Nations (FAO) (no date) 'Aquastat country database' [online]. <http://www.fao.org/nr/water/aquastat/data/query/index.html?lang=en> [accessed 12 September 2020].

FAO (2003) 'Review of world water resources by country', Water Reports No. 23, Rome. <http://www.fao.org/3/Y4473E/y4473e.pdf> [accessed 12 September 2020].

Fonseca, C., Franceys, R., Batchelor, C., McIntyre, P., Klutse, A., Komives, K., Moriarty, P., Naafs, A., Nyarko, K., Pezon, C., Potter, A., Reddy, R. and Snehalatha, M. (2010) 'Life-cycle costs approach: glossary and cost components', WASHCost Briefing note 1, IRC International Water and Sanitation Centre. <https://www.ircwash.org/sites/default/files/Fonseca-2010-Life.pdf> [accessed 18 September 2020].

Fonseca, C., Smits, S., Nyarko, K., Naafs, A. and Franceys, R. (2013) 'Financing capital maintenance of rural water supply systems: current practices and future options', WASHCost Working paper 9, IRC International Water and Sanitation Centre, March. <https://www.ircwash.org/sites/default/files/201303_9_wp_capmanex_web.pdf> [accessed 23 September 2020].

Foster, S., Chilton, J., Nijsten, G-J. and Richts, A. (2013) 'Groundwater – a global focus on the "local resource"', *Current Opinion in Environmental Sustainability* 5: 685–95 <https://doi.org/10.1016/j.cosust.2013.10.010>.

Foster, T. (2013) 'Predictors of sustainability for community-managed handpumps in sub-Saharan Africa: evidence from Liberia, Sierra Leone and Uganda', *Environmental Science and Technology* 47 (21): 12037–46 <https://doi.org/10.1021/es402086n>.

Foster, T. and Hope, R. (2016) 'A multi-decadal and social-ecological systems analysis of community waterpoint payment behaviours in rural Kenya', *Journal of Rural Studies* 47, 85–96 <https://doi.org/10.1016/j.jrurstud.2016.07.026>.

Foster, T. and Hope, R. (2017) 'Evaluating waterpoint sustainability and access implications of revenue collection approaches in rural Kenya', *Water Resources Research* 53 (2): 1473–90 <https://doi.org/10.1002/2016WR019634>.

Foster, T., Furey, S., Banks, B. and Willetts, J. (2019) 'Functionality of handpump water supplies: a review of data from sub-Saharan Africa and the Asia-Pacific region', *International Journal of Water Resources Development* 36 (5): 855–69 <https://doi.org/10.1080/07900627.2018.1543117>.

Foster, T., Shantz, A., Lala, S. and Willetts, J. (2018) 'Factors associated with operational sustainability of rural water supplies in Cambodia', *Environmental Science Water Research and Technology* 4: 1577–88 <https://doi.org/10.1039/C8EW00087E>.

Foster, T., Willetts, J., Lane, M., Thomson, P., Katuva, J. and Hope R. (2018) 'Risk factors associated with rural water supply failure: a 30-year retrospective study of handpumps on the south coast of Kenya', *Science of the Total Environment* 626: 156–64 <https://doi.org/10.1016/j.scitotenv.2017.12.302>.

Fox, C.J.E., Lopez-Alascio, B., Fay, M., Nicolas, C. and Rozenberg, J. (2019) 'Water, sanitation and irrigation', in J. Rozenberg and M. Fay (eds) (2019) *Beyond the Gap: How Countries Can Afford the Infrastructure they Need While Protecting the Planet*, pp. 47–70, Sustainable Infrastructure Series, Washington, DC: World Bank. <https://openknowledge.worldbank.org/handle/10986/31291> [accessed 23 September 2020].

Fraenkel, P. (1997) *Water Pumping Devices: A Handbook for Users and Choosers*, 2nd edn, Rugby: IT Publications Ltd.

Fraenkel, P. and Thake, J. (2006) *Water Lifting Devices: A Handbook for Users and Choosers*, 3rd edn, Rugby: Practical Action Publishing/FAO/It Power.

Franceys, R. (2005a) 'Charging to enter the water shop?' *Waterlines* 24 (2): 5–7. <https://www.ircwash.org/sites/default/files/Franceys-2005-Charging.pdf> [accessed 25 September 2020].

Franceys, R. (2005b) 'Charging to enter the water shop? The costs of urban water connections for the poor', *Water Supply* 5 (6): 209–16. <https://iwaponline.com/ws/article-pdf/5/6/209/417862/209.pdf> [accessed 25 September 2020].

Franceys, R. (2019) 'Utilitisation', Paper presented at the WASH Systems Symposium, The Hague, the Netherlands, 12–14 March. <https://www.ircwash.org/resources/'utilitisation'> [accessed 19 August 2020].

Franceys, R., Cavill, S. and Trevett, A. (2016) 'Who really pays? A critical overview of the practicalities of funding universal access', *Waterlines* 35 (1): 78–93 <https://doi.org/10.3362/1756-3488.2016.007>.

Freeze, R.A. and Cherry, J.A. (1979) *Groundwater*, Englewood Cliffs, NJ: Prentice-Hall. <http://hydrogeologistswithoutborders.org/wordpress/textbook-project/> [accessed 12 September 2020].

Geere, J-A. and Cortobius, M. (2017) 'Who carries the weight of water? Fetching water in rural and urban areas and the implications for water security', *Water Alternatives* 10 (2): 513–40 <http://www.water-alternatives.org/index.php/alldoc/articles/vol10/v10issue2/368-a10-2-18/file> [accessed 11 September 2020].

Geere, J-A., Bartram, J., Bates, L., Danquah, L., Evans, B., Fisher, M.B., Groce, N., Majuru, B., Mokoena, M.M., Mukhola, M.S., Nguyen-Viet, H., Duc, P.P., Williams, A.R., Schmidt, W-P. and Hunter, P.R. (2018) 'Carrying water may be a major contributor to disability from musculoskeletal disorders in low income countries: a cross-sectional survey in South Africa, Ghana and Vietnam', *Journal of Global Health* 8 (1): 1–14 <https://doi.org/10.7189/jogh.08.010406>.

Gerlach, E. (2019) 'Regulating rural water supply services: a comparative review of existing and emerging approaches with a focus on GIZ partner countries', GIZ Deutsche Gesellschaft für Internationale Zusammenarbeit GmbH, Eschborn, August. <https://www.giz.de/de/downloads/giz2019_Regulating_Rural_Water_Supply_Services.pdf> [accessed 16 July 2020].

Global Burden of Disease Collaborators (2019) 'Quantifying risks and interventions that have affected the burden of diarrhoea among children younger than 5 years: an analysis of the Global Burden of Disease Study 2017', *Lancet Infectious Diseases* 20 (1): 37–59 <https://doi.org/10.1016/S1473-3099(19)30401-3>.

Goldman, I. and Pabari, M. (2020) *Using Evidence in Policy and Practice: Lessons from Africa*, London: Routledge <https://doi.org/10.4324/9781003007043>.

Gosling, L. (2010) 'Equity and inclusion: a rights-based approach', WaterAid report, January.

Haanen, R. (2016) '130.000 rope pumps worldwide: 25 years experiences from Nicaragua, Ethiopia, Tanzania and six other countries', Paper presented at the 7th RWSN Forum, Abidjan, Côte d'Ivoire, 29 November – 2 December. <http://www.ropepumps.org/uploads/2/9/9/2/29929105/130.000_rope_pumps._r.h._paper_rwsn_2016.pdf> [accessed 17 September 2020].

Haider, H., Sadiq, R. and Tesfamariam, S. (2014) 'Performance indicators for small- and medium-sized water supply systems: a review', *Environmental Review* 22 (1): 1–40. <https://dx.doi.org/10.1139/er-2013-0013>.

Hamilton, K., Reyneke, B., Waso, M., Clement, T., Ndlovu, T., Khan, W., DiGiovanni, K., Rakestraw, E., Montalto, F., Haas, C. and Ahmed, W. (2019) 'A global review of the microbiological quality and potential health risks associated with roof-harvested rainwater tanks', *Clean Water* 2 (7) <https://doi.org/10.1038/s41545-019-0030-5> [accessed 12 September 2020].

Hannes, P., Hörtnagl, P., Reche, I. and Sommaruga, R. (2014) 'Bacterial diversity and composition during rain events with and without Saharan dust reaching a high mountain lake in the Alps', *Environmental Microbiology Reports* 6 (6): 618–24. <https://www.ncbi.nlm.nih.gov/pmc/articles/PMC4733657/> [accessed 12 September 2020].

Harvey, P.A. (2004) 'Borehole sustainability in rural Africa: an analysis of routine field data', paper presented at the 30th WEDC International Conference, Vientiane, Lao PDR, 25–29 October. <https://repository.lboro.ac.uk/articles/Borehole_sustainability_in_rural_Africa_an_analysis_of_routine_field_data/9594875> [accessed 20 September 2020].

Harvey, P.A. and Reed, R.A. (2007) 'Community-managed water supplies in Africa: sustainable or dispensable?', *Community Development Journal* 42 (3): 365–78. <https://www.ircwash.org/sites/default/files/Harvey-2007-Community.pdf> [accessed 26 September 2020].

Healy, R.W. (2013) *Estimating Groundwater Recharge*, Cambridge: Cambridge University Press <https://doi.org/10.1017/CBO9780511780745>.

Ho, J.C., Russel, K.C. and Davis, J. (2014) 'The challenge of global water access monitoring: evaluating straight-line distance versus self-reported travel time among rural households in Mozambique', *Journal of Water and Health* 12 (1): 173–83 <https://doi.org/10.2166/wh.2013.042>.

Hofkes, E.H. (ed.) *Small Community Water Supplies: Technology of Small Water Supply Systems in Developing Countries*, Chichester; New York, NY; Brisbane;

Toronto; Singapore: IRC / John Wiley and Sons. <https://www.ircwash.org/sites/default/files/201-83SM-2725.pdf> [accessed 20 September 2020].

Holtslag, H. (2020) 'Timeline: rural water supply', NICC. <http://stichtingnicc.nl/wp-content/uploads/2020/10/Timeline-Handpumps-NICC-Vrs.-19-10-2020_jm_b2.pdf> [accessed 4 November 2020].

Homoncik, S.C., MacDonald, A.M., Heal, K.V., Ó Dochartaigh, B.É. and Ngwenya, B.T. (2010) 'Manganese concentrations in Scottish groundwater', *Science of the Total Environment* 408, 2467–73 <https://doi.org/10.1016/j.scitotenv.2010.02.017>.

Hope, R. (2015) 'Is community management the community's choice? Implications for water and development policy in Africa', *Water Policy* 17: 664–78 <https://doi.org/10.2166/wp.2014.170>.

Hope, R., Foster, T., Koehler, J. and Thomson, P. (2019) 'Rural water policy in Africa and Asia', in S.J. Dadson, D.E. Garrick, E.C. Penning-Rowsell, J.W. Hall, R. Hope and J. Hughes (eds) *Water Science, Policy and Management: A Global Challenge*, pp. 159–79, Hoboken, NJ: John Wiley and Sons Ltd <https://onlinelibrary.wiley.com/doi/10.1002/9781119520627.ch9>.

House, S., Ferron, S., Sommer, M. and Cavill, S. (2014) 'Violence, gender and WASH: a practitioner's toolkit – Making water, sanitation and hygiene safer through improved programming and services', London: WaterAid/SHARE. <https://violence-wash.lboro.ac.uk> [accessed 10 September 2020].

Howard, G. and Bartram J. (2003) *Domestic Water Quantity, Service, Level and Health*, WHO: Geneva. <https://apps.who.int/iris/bitstream/handle/10665/67884/WHO_SDE_WSH_03.02.pdf?sequence=1&isAllowed=y> [accessed 11 September 2020].

Humphrey, J.H., Mbuya, M.N., Ntozini, R., Moulton, L.H., Stoltzfus, R.J., Tavengwa, N.V., Mutasa, K., Majo, F., Mutasa, B., Mangwadu, G., Chasokela, C.M., Chigumira, A., Chasekwa, B., Smith, L.E., Tielsch, J.M., Jones, A.D., Manges, A.R., Maluccio, J.A. and Prendergast, A.J. (2019) 'Independent and combined effects of improved water, sanitation, and hygiene, and improved complementary feeding, on child stunting and anaemia in rural Zimbabwe: a cluster-randomised trial', Lancet Global Health 7 (1): e132–47 <https://doi.org/10.1016/S2214-109X(18)30374-7>.

Hunter, P.R. (2009) 'Household water treatment in developing countries: comparing different intervention types using meta-regression', *Environmental Science and Technology* 43 (23): 8991–7 <https://doi.org/10.1021/es9028217>.

Hussein, M.K. and Muriaas, R.L. (2007) 'Traditional leaders in Malawi', in N. Patel and L. Svåsand (eds), *Government and Politics in Malawi*, pp. 155–73, Zomba: Kachere Series.

Huston, A. and Moriarty, P. (2018a) 'Building strong WASH systems for the SDGs: understanding the WASH system and its building blocks', IRC Working paper. <https://www.ircwash.org/sites/default/files/uploads/084-201813wp_buildingblocksdef_web.pdf> [accessed 9 July 2020].

Huston, A. and Moriarty, P. (2018b) 'Understanding the WASH system and its building blocks', IRC Working paper. <https://www.ircwash.org/sites/default/files/084-201813wp_buildingblocksdef_newweb.pdf> [accessed 4 September 2020].

Hutchings, P. and Carter, R.C. (2018) 'Setting SDG ambitions in a realistic time-scale', *Waterlines* 37 (1): 1–4 <https://doi.org/10.3362/1756-3488.2018.37-1ED>.

Hutchings, P., Chan, M.Y., Cuadrado, L., Ezbakhe, F., Mesa, B., Tamekawa, C. and Franceys, R. (2015) 'A systematic review of success factors in the community management of rural water supplies over the past 30 years', *Water Policy* 17 (5): 963–83 <https://doi.org/10.2166/wp.2015.128>.

Hutton, G. (2012a) 'Global costs and benefits of drinking water supply and sanitation interventions to reach the MDG target and universal coverage', WHO, WHO/HSE/WSH/12.01. <https://www.who.int/water_sanitation_health/publications/2012/globalcosts.pdf> [accessed 25 September 2020].

Hutton, G. (2012b) 'Monitoring "affordability" of water and sanitation services after 2015: review of global indicator options. A paper submitted to the United Nations Office of the High Commission for Human Rights' IRC WASH. <https://www.ircwash.org/resources/monitoring-"affordability"-water-and-sanitation-services-after-2015-review-global> [accessed 24 September 2020].

Hutton, G. and Varughese, M. (2016) 'The costs of meeting the 2030 Sustainable Development Goal targets on drinking water, sanitation, and hygiene', Water and Sanitation Program Working paper, World Bank, Washington DC. <https://openknowledge.worldbank.org/handle/10986/23681> [accessed 23 September 2020].

Hydroconseil/Trémolet Consulting (2015) 'End-of-project evaluation: Triple-S water services that last, an IRC initiative 2009–2014: main report', final version, January. <https://www.ircwash.org/resources/end-project-evaluation-epe-triple-s---water-services-last-irc-initiative-2009-2014-main> [accessed 19 August 2020].

Intergovernmental Panel on Climate Change (IPCC) (2020) Climate Change and Land: An IPCC Special Report on Climate Change, Desertification, Land Degradation, Sustainable Land Management, Food Security, and Greenhouse Gas Fluxes in Terrestrial Ecosystems –Summary for Policymakers. <https://www.ipcc.ch/site/assets/uploads/sites/4/2020/02/SPM_Updated-Jan20.pdf> [accessed 12 September 2020].

International Association of Hydrogeologists (IAH) (2015) 'Food security and groundwater', Strategic Overview Series, <https://iah.org/wp-content/uploads/2015/11/IAH-Food-Security-Groundwater-Nov-2015.pdf> [accessed 12 September 2020].

The International Benchmarking Network (IBNET) (no date) 'IBNET Indicators'. <https://www.ib-net.org/toolkit/ibnet-indicators/> [accessed 23 July 2020].

International Development Partners Group, Nepal (IDPG) (2017) 'A common framework for gender equality and social inclusion', Gender Equality and Social Inclusion Working Group. <https://www.undp.org/content/dam/nepal/docs/generic/GESI%20framework%20Report_Final_2017.pdf> [accessed 26 September 2020].

IRC (2019) 'Taking a systems strengthening approach' [video], 23 April. <https://www.ircwash.org/resources/taking-systems-strengthening-approach> [accessed 29 September 2020].

Jarvis, N., Koestel, J. and Larsbo, M. (2016) 'Understanding preferential flow in the vadose zone: recent advances and future prospects', *Vadose Zone Journal* 15 (12): 1–11 <https://doi.org/10.2136/vzj2016.09.0075>.

Jimenez-Redal, R., Soriano, J., Holowko, N., Almandoz, J. and Arregui, F. (2017) 'Assessing sustainability of rural gravity-fed water schemes on Idjwi

Island, D.R. Congo', *International Journal of Water Resources Development* 34 (6): 1022–35 <https://doi.org/10.1080/07900627.2017.1347086>.

Jiménez, A., Kjellén, M. and Le Deunff, H. (2015) 'Accountability in WASH: explaining the concept', Accountability for Sustainability Partnership: UNDP Water Governance Facility at SIWI and UNICEF. <https://www.unicef.org/wash/files/Accountability_in_WASH_Explaining_the_Concept.pdf> [accessed 9 July 2020].

Jiménez, A., LeDeunff, H., Giné, R., Sjödin, J., Cronk, R., Murad, S., Takane, M. and Bartram, J. (2019) 'The enabling environment for participation in water and sanitation: a conceptual framework', *Water* 11 (2): 308 <https://doi.org/10.3390/w11020308>.

Joint Monitoring Programme (JMP) (2000) *Global Water Supply and Sanitation Assessment: 2000 Report*, WHO/UNICEF <https://www.who.int/water_sanitation_health/monitoring/jmp2000.pdf> [accessed 12 September 2020].

JMP (2017) *Progress on Drinking Water, Sanitation and Hygiene: 2017 Update And SDG Baselines.* https://www.who.int/water_sanitation_health/publications/jmp-2017/en/ [accessed 4 August 2020].

JMP (2018a) 'Core questions' [online], 2018 update, UNICEF and WHO. <https://washdata.org/monitoring/methods/core-questions> [accessed 11 September 2020].

JMP (2018b) 'Schools' [online], UNICEF and WHO <https://washdata.org/monitoring/schools> [accessed 11 September 2020].

JMP (2019a) *Progress on Household Drinking Water, Sanitation and Hygiene, 2000-2017: Special Focus on Inequalities*, New York, NY: UNICEF and WHO. <https://washdata.org/sites/default/files/documents/reports/2019-07/jmp-2019-wash-households.pdf> [accessed 23 July 2020].

JMP (2019b) *WASH in Health Care Facilities: Global Baseline Report 2019*, Geneva: UNICEF and WHO. <https://washdata.org/monitoring/health-care-facilities> [accessed 24 September 2020].

JMP (2020a) *Progress on Drinking Water, Sanitation and Hygiene in Schools: Special Focus on COVID-19*, New York, NY: UNICEF and WHO. <https://washdata.org/sites/default/files/2020-08/jmp-2020-wash-schools.pdf> [accessed 24 September 2020].

JMP (2020b) 'Data' [online]. <https://washdata.org/data> [accessed July 2020].

Jones, H. and Wilbur, J. (2014) 'Compendium of accessible WASH technologies', WEDC/WaterAid/Share. <www.wateraid.org/accessibleWASHtechnologies> [accessed 26 September 2020].

Jordan, T.D. (1984) *A Handbook of Gravity-Flow Water Systems*, London: IT Publications Ltd.

Joshi, A. and Carter, B. (2015) *Public Sector Institutional Reform: Topic Guide*, Birmingham: GSDRC, University of Birmingham. <https://gsdrc.org/wp-content/uploads/2015/07/PSIR_TG.pdf> [accessed 16 July 2020].

Kebede, S., Fallas, H.C., MacAllister, D.J., Dessie, N., Tayitu, Y., Kefale, Z., Wolde, G., Whaley, L., Banks, E., Casey, V. and MacDonald, A.M. (2019) 'UPGro Hidden Crisis Research consortium: technical brief – Ethiopia', British Geological Survey open report, OR/19/055. <http://nora.nerc.ac.uk/id/eprint/527020/1/OR19055.pdf> [accessed 17 September 2020].

Kebede, S., MacDonald, A.M., Bonsor, H.C., Dessie, N., Yehualaeshet, T., Wolde, G., Wilson, P., Whaley, L. and Lark, R.M. (2017) 'UPGro Hidden Crisis Research consortium: unravelling past failures for future success in rural water supply. Survey 1 results, country report Ethiopia', British Geological Survey open report, OR/17/024. <http://nora.nerc.ac.uk/id/eprint/516998/> [accessed 17 September 2020].

Kelley, M., Ferrand, R.A., Muraya, K., Chigudu, S., Molyneux, S., Pai, M. and Barasa, E. (2020) 'An appeal for practical social justice in the COVID-19 global response in low-income and middle-income countries', *The Lancet* 8 (7): E888–9 <https://doi.org/10.1016/S2214-109X(20)30249-7>.

King, M.L.Jr. (1967) 'Where do we go from here?', Address delivered at the Eleventh Annual SCLC Convention, Atlanta, GA, 16 August <https://kinginstitute.stanford.edu/king-papers/documents/where-do-we-go-here-address-delivered-eleventh-annual-sclc-convention> [accessed 7 September 2020].

Kleemeier, E. (2000) 'The impact of participation on sustainability: an analysis of the Malawi rural piped scheme program', *World Development* 28 (5): 929–44 <https://doi.org/10.1016/S0305-750X(99)00155-2>.

Kleemeier, E. (2001) 'The role of government in maintaining rural water supplies: caveats from Malawi's gravity schemes', *Public Administration and Development* 21 (3): 245–57 <https://doi.org/10.1002/pad.171>.

Kleemeier, E. and Narkevic, J. (2010) 'Private operator models for community water supply: a global review of private operator experiences in rural areas', Water and Sanitation Program Field note, World Bank, Washington, DC, February. <https://www.wsp.org/sites/wsp.org/files/publications/Private_Operator ModelsforCommunity_WaterSupply.pdf> [accessed 20 September 2020].

Konikow, L.F. (2011) 'Contribution of global groundwater depletion since 1900 to sea-level rise', *Geophysical Research Letters* 38, L17401 <https://doi.org/10.1029/2011GL048604>.

Korpe, P.S. and Petri, W.A. (2012) 'Environmental enteropathy: critical implications of a poorly understood condition,' *Trends in Molecular Medicine* 18 (6): 328–36 <https://doi.org/10.1016/j.molmed.2012.04.007>.

Kremer, M., Mullainathan, S., Zwane, A., Miguel, E. and Null, C. (2011) 'Social engineering: evidence from a suite of take-up experiments in Kenya', unpublished draft <https://www.poverty-action.org/sites/default/files/publications/chlorinedispensers.pdf> [accessed 10 September 2020].

Kruseman, G.P. and de Ridder, N.A. (1994) *Analysis and Evaluation of Pumping Test Data*, 2nd edn, Wageningen: International Institute for Land Reclamation and Improvement. <https://www.hydrology.nl/images/docs/dutch/key/Kruseman_and_De_Ridder_2000.pdf> [accessed 12 September 2020].

Kummu, M., de Moel, H., Ward, P.J. and Varis, O. (2011) 'How close do we live to water? A global analysis of population distance to freshwater bodies', *PLoS ONE* 6 (6): e20578 <https://doi.org/10.1371/journal.pone.0020578>.

Langenegger, O. (1994) 'Groundwater quality and handpump corrosion in Africa', UNDP-World Bank Water and Sanitation Program, Water and Sanitation report 8. <http://documents1.worldbank.org/curated/en/998791468325792705/pdf/multi-page.pdf> [accessed 16 September 2020].

Langford, M. (2017). 'Socio-economic rights', in R. Maliks and J. Schaffer (eds.) *Moral and Political Conceptions of Human Rights: Implications for Theory and*

Practice, pp. 258–98, Cambridge: Cambridge University Press <https://doi.org/10.1017/9781316650134.013>.

Lapworth, D.J., Krishnan, G., MacDonald, A.M. and Rao, M.S. (2017) 'Groundwater quality in the alluvial aquifer system of northwest India: new evidence of the extent of anthropogenic and geogenic contamination', *Science of the Total Environment* 599 (600): 1433–44 <https://doi.org/10.1016/j.scitotenv.2017.04.223>.

Lapworth, D.J., MacDonald, A.M., Kebede, S., Owor, M., Chavula, G., Fallas, H., Wilson, P., Ward, J.S.T., Lark, M., Okullo, J., Mwathunga, E., Banda, S., Gwengweya, G., Nedaw, D., Jumbo, S., Banks, E., Cook, P. and Casey, V. (2020) 'Drinking water quality from rural handpump-boreholes in Africa', *Environmental Research Letters* 15 (6): 064020 <https://doi.org/10.1088/1748-9326/ab8031>.

Le Sève, M.D. (2018) 'A political economy analysis of Uganda's rural water supply sector', ODI Report, October. <https://upgrohiddencrisisdotorg.files.wordpress.com/2019/08/pea_odi_uganda.pdf> [accessed 17 September 2020].

Lerner, D.N., Issar, A.S. and Simmers, I. (1990) Groundwater Recharge: A Guide to Understanding and Estimating Natural Recharge, Hannover: Heinz Heise.

Liddle, E, Mager, S.M. and Nel, E.L. (2014) 'The importance of community-based informal water supply systems in the developing world and the need for formal sector support', *Geographical Journal* 182 (1): 85–96 <https://doi.org/10.1111/geoj.12117>.

Liddle, E. and Fenner, R. (2017). 'Water point failure in sub-Saharan Africa: the value of a systems thinking approach', *Waterlines* 36 (2):140–66 <https://doi.org/10.3362/1756-3488.16-00022>.

Lockwood, H. (2019) 'Sustaining rural water: a comparative study of maintenance models for community managed schemes', Sustainable WASH Systems Learning Partnership, Research Report, USAID, July. <https://www.ircwash.org/sites/default/files/sustaining_rural_water_-_a_comparative_study_of_maintenance_models_for_community-managed_schemes.pdf> [accessed 20 September 2020].

Lockwood, H. and Le Gouais, H. (2015) 'Professionalising community-based management for rural water services', Briefing note, IRC Building blocks for sustainability series, March. <https://www.ircwash.org/sites/default/files/084-201502triple-s_bn01defweb_1_0.pdf> [accessed 16 July 2020].

Lockwood, H., Casey, V. and Tillett, W. (2018) 'Management models for piped water supply services: a decision resource for rural and small-town contexts', Aguaconsult/WaterAid. <https://washmatters.wateraid.org/sites/g/files/jkxoof256/files/management-models-for-piped-water-supply-services.pdf> [accessed 11 November 2020].

Luby, S.P., Rahman, M., Arnold, B.F., Unicomb, L., Ashraf, S., Winch, P.J., Stewart, C.P., Begum, F., Hussain, F., Benjamin-Chung, J., Leontsini, E., Naser, A.M., Parvez, S.M., Hubbard, A.E., Lin, A., Nizame, F.A., Jannat, K., Ercumen, A., Ram, P.K., Das, K.K., Abedin, J., Clasen, T.F., Dewey, K.G., Fernald, L.C., Null, C., Ahmed, T. and Colford, J.M. (2018) 'Effects of water quality, sanitation, handwashing, and nutritional interventions on diarrhoea and child growth in rural Bangladesh: a cluster randomised controlled trial', Lancet Global Health 6 (3): e302–15 <http://dx.doi.org/10.1016/S2214-109X(17)30490-4>.

Luseka, E. (2020) 'Initiating de-colonisation of WASH sector knowledge' [blogpost], *Medium*, 8 June. <https://medium.com/@euphresia_luseka/initiating-de-colonization-of-wash-sector-knowledge-c8ad0a9f8d6> [accessed 3 December 2020].

MacArthur, J. (2015) 'Handpump standardisation in sub-Saharan Africa: seeking a champion', Publication 2015-1, RWSN. <http://www.rural-water-supply.net/_ressources/documents/default/1-652-2-1421834932.pdf> [accessed 15 September 2020].

MacCarthy M.F., Buckingham J.W. and Mihelcic J.R. (2013) 'Increasing access to low-cost water supplies in rural areas: EMAS household water supply technologies in Bolivia', RWSN Field note 2013-4. <https://www.emas-international.de/images/weltweit/Fieldnote_2013-04_01.pdf> [accessed 4 November 2020].

MacDonald A.M., Bell R.A., Kebede S., Azagegn T., Yehualaeshet T., Pichon F., Young M., McKenzie A.A., Lapworth D.J., Black E., Calow R.C. (2019) 'Groundwater and resilience to drought in the Ethiopian highlands', *Environmental Research Letters* 14 (9): 095003 <https://doi.org/10.1088/1748-9326/ab282f>.

MacDonald, A.M., Bonsor, H.C., Ó Dochartaigh, B.É. and Taylor, R.G. (2012) 'Quantitative maps of groundwater resources in Africa', *Environmental Research Letters* 7 (2): 024009. <https://iopscience.iop.org/article/10.1088/1748-9326/7/2/024009> [accessed 12 September 2020].

MacDonald, A.M, Davies, J., Calow, R. and Chilton, J. (2005) *Developing Groundwater: A Guide for Rural Water Supply*, Rugby: ITDG Publishing. <http://practicalactionpublishing.com/book/489/developing-groundwater> [accessed 12 September 2020].

Margat, J. and van der Gun, J. (2013) *Groundwater Around the World: A Geographic Synopsis*, Boca Raton, FL: CRC Press.

Marks, S.J., Komives, K. and Davis, J. (2014) 'Community participation and water supply sustainability: evidence from handpump projects in rural Ghana', *Journal of Planning Education and Research* 34 (3): 276–86 <https://doi.org/10.1177/0739456X14527620>.

McKay, A. and Deshingkar, P. (2014) 'Internal remittances and poverty: further evidence from Africa and Asia', Migrating Out of Poverty Research Consortium, Working paper 12, March. <https://assets.publishing.service.gov.uk/media/57a089c6e5274a31e000026a/WP12-migratingpov.pdf> [accessed 21 May 2020].

McNicholl, D., Hope, R., Money, A., Lane, A., Armstrong, A., Dupuis, M., Harvey, A., Nyaga, C., Womble, S., Allen, J., Katuva, J., Barbotte, T., Lambert, L., Staub, M., Thomson, P. and Koehler, J. (2020) 'Results-based contracts for rural water services', Uptime Consortium Working paper 2. <https://www.rural-water-supply.net/fr/ressources/details/931> [accessed 27 September 2020].

McNicholl, D., Hope, R., Money, A., Lane, A., Armstrong, A., van der Wilk, N., Dupuis, M., Harvey, A., Nyaga, C., Womble, S., Favre, D., Allen, J., Katuva, J., Barbotte, T., Buhungiro, E., Thomson, P. and Koehler, J. (2019) 'Performance-based funding for reliable rural water services in Africa', Uptime Consortium

Working paper 1. <https://www.smithschool.ox.ac.uk/research/water/report-performance-based-funding.html> [accessed 7 July 2020].

Meadows, D.H. (2008) *Thinking in Systems: A Primer*, Sterling, VA: Earthscan. <https://research.fit.edu/media/site-specific/researchfitedu/coast-climate-adaptation-library/climate-communications/psychology-amp-behavior/Meadows-2008.-Thinking-in-Systems.pdf> [accessed 4 September 2020].

Meera, V. and Ahammed, M.M. (2006) 'Water quality of rooftop rainwater harvesting systems: a review', *Journal of Water Supply: Research and Technology–AQUA* 55 (4): 257–68 <https://doi.org/10.2166/AQUA.2006.0010>.

Minasny, B., Fiantis, D., Mulyanto, B., Sulaeman, Y. and Widyatmanti, W. (2020) 'Global soil science research collaboration in the 21st century: time to end helicopter research', *Geoderma* 373, 114302 <https://doi.org/10.1016/j.geoderma.2020.114299>.

Ministry of Water and Environment (Uganda) (2012) 'Steps in implimentation [sic] of water and sanitation software activities: a guide to district local governments', Directorate of Water Development, July. <https://www.mwe.go.ug/sites/default/files/library/Software%20Steps%202012%20final%20booklet.pdf> [accessed 21 May 2020].

Misstear, B., Banks, D. and Clark, L. (2017) Water Wells and Boreholes, 2nd edn, Chichester: Wiley Blackwell.

Mitchell, T.D. and Hulme M. (1999) 'Predicting regional climate change: living with uncertainty', *Progress in Physical Geography* 23 (1): 57–78. <https://crudata.uea.ac.uk/~timm/papers/pipg-99.pdf> [accessed 12 September 2020].

Mohapatra, M., Anand, S., Mishra, B.K., Giles, D.E. and Singh, P. (2009) 'Review of fluoride removal from drinking water', *Journal of Environmental Management* 91 (1): 67–77 <https://doi.org/10.1016/j.jenvman.2009.08.015>.

Morgan, P. (1990) Rural Water Supplies and Sanitation: A Text from Zimbabwe's Blair Research Laboratory, London: MacMillan.

Moriarty, P., Smits, S., Butterworth, J. and Franceys, R. (2013) 'Trends in rural water supply: towards a service delivery approach', *Water Alternatives* 6 (3): 329–49. <http://www.water-alternatives.org/index.php/volume6/v6issue3/220-a6-3-1/file> [accessed 7 July 2020].

Morse, S.S., Mazet, J.A.K., Woolhouse, M., Parrish, C.R., Carroll, D., Karesh, W.B., Zambrana-Torrelio, C., Lipkin, W.I, and Daszak, P. (2012) 'Prediction and prevention of the next pandemic zoonosis', *The Lancet* 380: 1956–65. <https://www.thelancet.com/action/showPdf?pii=S0140-6736%2812%2961684-5> [accessed 14 July 2020].

Mudgal, A.K. (1997) 'India handpump revolution: challenge and change', HTN Working paper, WP 01/97. <https://www.rural-water-supply.net/_ressources/documents/default/1-436-2-1355213186.pdf> [accessed 17 September 2020].

Mwathunga, E., Fallas, H.C., MacAllister, D.J., Mkandawire, T., Makuluni, P., Shaba, C., Jumbo, S., Moses, D., Whaley, L., Banks, E., Casey, V. and MacDonald, A.M. (2019) 'Physical factors contributing to rural water supply functionality performance in Malawi', BGS Open report, OR/19/057. <http://nora.nerc.ac.uk/id/eprint/527013/> [accessed 17 September 2020].

Mwathunga, E., MacDonald, A.M., Bonsor, H.C., Chavula, G., Banda, S., Mleta, P., Jumbo, S., Gwengweya, G., Ward, J., Lapworth, D., Whaley, L. and Lark, R.M. (2017) 'UPGro Hidden Crisis Research consortium. Survey 1 country report, Malawi', BGS Open report, OR/17/046. <http://nora.nerc.ac.uk/id/eprint/518402/> [accessed 17 September 2020].

Nafisi, A. (2005) 'Mysterious connections that link us together', This I Believe [radio programme], NPR, Morning Edition, 18 July. <https://www.npr.org/templates/story/story.php?storyId=4753976&t=1601374448747> [accessed 29 September 2020].

Narayan, D. and Petesch, P. (2002) *Voices of the Poor: From Many Lands*, Volume 3, Oxford: Oxford University Press for the World Bank. <http://documents.worldbank.org/curated/en/864691468233712217/Voices-of-the-poor-from-many-lands> [accessed 14 July 2020].

Narayan, D., Chambers, R., Shah, M. and Petesch, P. (2000) *Voices of the Poor: Crying Out for Change*, Volume 2, Oxford: Oxford University Press for the World Bank. <http://documents.worldbank.org/curated/en/501121468325204794/Voices-of-the-poor-crying-out-for-change> [accessed 14 July 2020].

Narayan, D., Patel, R., Schafft, K., Rademacher, A. and Koch-Shulte, S (2000) *Voices of the Poor: Can Anyone Hear Us?* Volume 1, Oxford: Oxford University Press for the World Bank. <http://documents.worldbank.org/curated/en/131441468779067441/Voices-of-the-poor-can-anyone-hear-us> [accessed 14 July 2020].

Neely, K. (ed.) (2019) *Systems Thinking and WASH: Tools and Case Studies for a Sustainable Water Supply*, Rugby: Practical Action Publishing. <https://www.pseau.org/outils/ouvrages/practical_action_systems_thinking_and_wash_tools_and_case_studies_for_a_sustainable_water_supply_2019.pdf> [accessed 20 August 2020].

Neier, A. (2006) 'Social and economic rights: a critique', *Human Rights Brief* 13 (2): 1–3. <https://digitalcommons.wcl.american.edu/cgi/viewcontent.cgi?article=1254&context=hrbrief> [accessed 15 July 2020].

Ngai, T.K.K., Shrestha, R.R., Dangol, B., Maharjan, M. and Murcott, S.E. (2007) 'Design for sustainable development – household drinking water filter for arsenic and pathogen treatment in Nepal', *Journal of Environmental Science and Health* 42 (12): 1879–88 <https://doi.org/10.1080/10934520701567148>.

Ngure, F.M., Humphrey, J.H., Mbuya, M.N.N., Majo, F., Mutasa, K., Govha, M., Mazarura, E., Chasekwa, B., Prendergast, A.J., Curtis, V., Boor, K.J. and Stoltzfus, R.J. (2013) 'Formative research on hygiene behaviors and geophagy among infants and young children and implications of exposure to fecal bacteria', *American Journal of Tropical Medicine and Hygiene* 89 (4): 709–16 <https://doi.org/10.4269/ajtmh.12-0568>.

Nicomel, N.R., Leus, K., Folens, K., van der Voort, P. and Du Laing, G. (2016) 'Technologies for arsenic removal from water: current status and future perspectives', *International Journal of Environmental Research and Public Health* 13 (1): 62 <https://doi.org/10.3390/ijerph13010062>.

Nowicki, S., Lapworth, D.J., Ward, J.S.T., Thomson, P. and Charles, K. (2019) 'Tryptophan-like fluorescence as a measure of microbial contamination risk in groundwater', *Science of The Total Environment* 646: 782–91 <https://doi.org/10.1016/j.scitotenv.2018.07.274>.

Null, C., Stewart, C.P., Pickering, A.J., Dentz, H.N., Arnold, B.F., Arnold, C.D., Benjamin-Chung, J., Clasen, T., Dewey, K.G., Fernald, L.C.H, Hubbard, A.E., Kariger, P., Lin, A., Luby, S.P., Mertens, A., Njenga, S.M., Nyambane, G.,

Ram, P.K. and Colford, J.M. (2018) 'Effects of water quality, sanitation, handwashing, and nutritional interventions on diarrhoea and child growth in rural Kenya: a cluster-randomised controlled trial', *The Lancet Global Health* 6 (3): e316–29 <https://doi.org/10.1016/S2214-109X(18)30005-6>.

Oates, N. and Mwathunga, E. (2018) 'A political economy analysis of Malawi's rural water supply sector', ODI Report, October. <https://upgrohiddencrisisdotorg.files.wordpress.com/2019/08/pea_odi_malawi.pdf> [accessed 16 September 2020].

Office of the United Nations High Commissioner for Human Rights (OHCHR) (no date) 'International covenant on economic, social and cultural rights'. <https://www.ohchr.org/en/professionalinterest/pages/cescr.aspx> [accessed 29 September 2020].

Ofwat (2015) 'Affordability and debt 2014–15', The Water Services Regulation Authority, December. <https://www.ofwat.gov.uk/wp-content/uploads/2015/12/prs_web20151201affordability.pdf> [accessed 24 September 2020].

Ohno, K., Yanase, T., Matsuo, Y., Kimura, T., Rahman, M.H., Magara, Y. and Matsui, Y. (2007) 'Arsenic intake via water and food by a population living in an arsenic-affected area of Bangladesh', *Science of The Total Environment* 381 (1): 68–76 <https://doi.org/10.1016/j.scitotenv.2007.03.019>.

Okidegbe, N. (2001) 'Rural poverty trends and measurement', Rural Development Strategy Background paper No. 3, World Bank, August. <http://documents.worldbank.org/curated/en/737021468769199842/pdf/multi0page.pdf> [accessed 19 May 2020].

Okri, B. (2011) 'Magical Ben Okri casts a spell on his readers' [interview], *CNN*, 28 June. <http://edition.cnn.com/2011/WORLD/africa/06/28/ben.okri.nigeria/index.html> [accessed 28 September 2020].

Olaerts, L., Walters, J.P., Linden, K.G., Javernick-Will, A. and Harvey, A. (2019) 'Factors influencing revenue collection for preventative maintenance of community water systems: a fuzzy-set qualitative comparative analysis', *Sustainability* 11 (13): 3726 <https://doi.org/10.3390/su11133726>.

Organisation for Economic Co-operation and Development (OECD) (2005) *Paris Declaration on Aid Effectiveness*, Paris: OECD Publishing <http://dx.doi.org/10.1787/9789264098084-en>.

OECD (2008) *Accra Agenda for Action*, Paris: OECD Publishing. <https://dx.doi.org/10.1787/9789264098107-en>.

OECD (2009) 'Managing water for all: an OECD perspective on pricing and financing. Key messages for policy makers'. <https://www.oecd.org/env/42350563.pdf> [accessed 25 September 2020].

OECD (2018) 'OECD-DAC blended finance principles for unlocking commercial finance for the sustainable development goals'. <https://www.oecd.org/dac/financing-sustainable-development/development-finance-topics/OECD-Blended-Finance-Principles.pdf> [accessed 19 August 2020].

Ostrom, E. (2007) 'Institutional rational choice: an assessment of the institutional analysis and development framework', in P.A. Sabatier (ed.) *Theories of the Policy Process*, pp. 21–64, Boulder, CO: Westview Press. <http://edwardwimberley.com/courses/IntroEnvPol/theorypolprocess.pdf> [accessed 18 September 2020].

Owor, M., Fallas, H.C., MacAllister, D.J., Okullo, J., Katusiime, F., Berochan, G., Whaley, L., Banks, E., Casey, V. and MacDonald, A.M. (2019) 'Physical factors contributing to rural water supply functionality performance in Uganda', British Geological Survey Open report, OR/19/056. <http://nora.nerc.ac.uk/id/eprint/527019/> [accessed 17 September 2020].

Owor, M., MacDonald, A.M., Bonsor, H.C., Okullo, J., Katusiime, F., Alupo, G., Berochan, G., Tumusiime, C., Lapworth, D., Whaley, L. and Lark, R.M. (2017) 'UPGro Hidden Crisis Research consortium. Survey 1 country report, Uganda', British Geological Survey Open report, OR/17/029. <http://nora.nerc.ac.uk/id/eprint/518403/> [accessed 17 September 2020].

Oxford Poverty and Human Development Initiative (OPHI) (2020) 'Global MPI data tables' [online]. <https://ophi.org.uk/multidimensional-poverty-index/data-tables-do-files/> [accessed 28 September 2020].

Pankhurst, A. (2008) 'The emergence, evolution and transformations of *iddir* funeral associations in urban Ethiopia', *Journal of Ethiopian Studies* 41 (1/2): 143–85 <https://www.jstor.org/stable/41967613>.

Parker, A., Cruddas, P., Rowe, N., Carter, R.C. and Webster, J. (2013) 'Tank costs for domestic rainwater harvesting in East Africa', Proceedings of the Institution of Civil Engineers, *Water Management* 166 (10): 536–45 <http://dx.doi.org/10.1680/wama.11.00113>.

Penakalapati, G., Swarthout, J., Delahoy, M.J., McAliley, L., Wodnik, B., Levy, K. and Freeman, M.C. (2017) 'Exposure to animal feces and human health: a systematic review and proposed research priorities', *Environmental Science & Technology* 51 (20): 11537–52 <https://doi.org/10.1021/acs.est.7b02811>.

Peters, T.J. and Waterman, R.H.Jr (1982) *In Search of Excellence: Lessons from America's Best-Run Companies*, London: Harper and Row.

Pichon, F. (2019) 'Rural water supply in Ethiopia: a political economy analysis', ODI Report, July. <https://upgrohiddencrisisdotorg.files.wordpress.com/2019/08/pea_odi_ethiopia.pdf> [accessed 14 September 2020].

Pickering, A.J. and Davis, J. (2012) 'Freshwater availability and water fetching distance affect child health in sub-Saharan Africa', *Environmental Science & Technology* 46 (4): 2391–7 <https://doi.org/10.1021/es203177v>.

Pond, K., King, R., Herschan, J., Malcolm, R., McKeown, R.M. and Schmoll, O. (2020) 'Improving risk assessments by sanitary inspection for small drinking-water supplies – qualitative evidence', *Resources* 9 (6): 71 <https://doi.org/10.3390/resources9060071>.

Post, D., Agarwal, S. and Venugopal, V. (2014) 'Rapid feedback: the role of community scorecards in improving service delivery', How-to-notes 88497, World Bank. <http://documents1.worldbank.org/curated/en/462221468333561977/pdf/884970WP0Rapid00Box385225B00PUBLIC0.pdf> [accessed 9 July 2020].

Price, M. (1996) *Introducing Groundwater*, 2nd edn, London and New York, NY: Taylor and Francis.

Priestley, S. and Rutherford, T. (2016) 'Water bills: affordability and support for household customers', Research Briefing paper CBP06596, House of Commons Library, 9 August. <https://commonslibrary.parliament.uk/research-briefings/sn06596/> [accessed 14 July 2020].

Public Health England (2020) 'Beyond the data: understanding the impact of COVID-19 on BAME groups'. <https://assets.publishing.service.gov.uk/government/uploads/system/uploads/attachment_data/file/892376/COVID_stakeholder_engagement_synthesis_beyond_the_data.pdf> [accessed 15 July 2020].

Ramalingham, B. and Jones, H. (2008) 'Exploring the science of complexity: ideas and implications for development and humanitarian efforts', 2nd edn, ODI Working paper 285. <https://www.odi.org/sites/odi.org.uk/files/odi-assets/publications-opinion-files/833.pdf> [accessed 20 August 2020].

Ravenscroft, P., Brammer, H. and Richards K. (2009) *Arsenic Pollution: A Global Synthesis*, Oxford: Blackwell.

Ravenscroft, P., Burgess, W.G., Ahmed, K.M., Burren, M. and Perrin, J. (2005) 'Arsenic in groundwater of the Bengal Basin, Bangladesh: distribution, field relations, and hydrogeological setting', *Hydrogeology Journal* 13, 727–51 <https://doi.org/10.1007/s10040-003-0314-0>.

Reygadas, F., Gruber, J.S., Dreizler, L., Nelson, K.L. and Ray I. (2018) 'Measuring user compliance and cost effectiveness of safe drinking water programs: a cluster-randomized study of household ultraviolet disinfection in rural Mexico', *American Journal of Tropical Medicine and Hygiene* 98 (3): 824–34 <https://doi.org/10.4269/ajtmh.17-0440>.

Reynolds, J. (1992) *Handpumps: Towards a Sustainable Technology. Research and Development During the Water Supply and Sanitation Decade*, Washington, DC: UNDP / World Bank. <https://www.ircwash.org/sites/default/files/232.2-92HA-10210.pdf> [accessed 17 September 2020].

Rihani, S. and Geyer, R. (2001) 'Complexity: an appropriate framework for development?' *Progress in Development Studies* 1 (3): 237–45. <https://www.researchgate.net/publication/254966739_Complexity_An_Appropriate_Framework_for_Development> [accessed 20 August 2020].

Rittel, H.W.J. and Webber, M.M. (1973) 'Dilemmas in a general theory of planning', *Policy Sciences* 4 (2): 155–69 <https://doi.org/10.1007/BF01405730>.

Roche, R., Bain, R. and Cumming, O. (2017) 'A long way to go – estimates of combined water, sanitation and hygiene coverage for 25 sub-Saharan African countries', *PLoS ONE* 12 (2): e0171783 <https://doi.org/10.1371/journal.pone.0171783>.

Rosa G. and Clasen T. (2010) 'Estimating the scope of household water treatment in low- and medium-income countries', *American Journal of Tropical Medicine and Hygiene* 82 (2): 289–300 <https://doi.org/10.4269/ajtmh.2010.09-0382>.

Rowles, R. (1995) *Drilling for Water: A Practical Manual*, 2nd edn, London: Ashgate Publishing.

Ruan, S. (2020) 'Likelihood of survival of coronavirus disease 2019', *Lancet Infectious Diseases* 20 (6): 630–1 <https://doi.org/10.1016/S1473-3099(20)30257-7>.

Rural Water Supply Network (RWSN) (no date, a) 'Hand-dug wells' [online]. <https://www.rural-water-supply.net/en/resources/sort/download_hits-desc/filter/2_217> [accessed 7 April 2020].

RWSN (no date, b) 'Handpump technologies' [online]. <https://www.rural-water-supply.net/en/implementation/public-domain-handpumps> [accessed 17 September 2020].

RWSN (no date, c) 'Harnessing the rain: rainwater harvesting' [online]. <https://www.rural-water-supply.net/en/self-supply/rainwater-harvesting> [accessed 7 April 2020].

RWSN (no date, d) 'Household water treatment' [online]. <http://www.rural-water-supply.net/en/search?search=household+water+treatment> [accessed 7 April 2020].

RWSN (no date, e) 'Manual drilling' [online]. <https://www.rural-water-supply.net/en/resources/sort/download_hits-desc/filter/2_145> [accessed 7 April 2020].

RWSN (no date, f) 'Professional drilling' [online]. <https://www.rural-water-supply.net/en/resources/sort/download_hits-desc/filter/2_32> [accessed 7 April 2020].

RWSN (no date, g) 'Professional water well drilling' [online]. <https://www.rural-water-supply.net/en/sustainable-groundwater-management/professionnal-water-well-drilling> [accessed 16 July 2020].

RWSN (no date, h) 'Self-supply' [online]. <https://www.rural-water-supply.net/en/self-supply> [accessed 7 July 2020].

RWSN (2009) 'Handpump data, selected countries in sub-Saharan Africa' [online]. <https://www.rural-water-supply.net/_ressources/documents/default/203.pdf> [accessed 17 September 2020].

RWSN (2010) 'Myths of the rural water supply sector', Perspectives No. 4, RWSN Executive Steering Committee, May. <https://www.rural-water-supply.net/_ressources/documents/default/226.pdf> [accessed 17 September 2020].

RWSN (2019) 'The RWSN International Directory of rural water supply models, tariffs and lifecycle costs', Skat Foundation, St Gallen, Switzerland. <https://www.rural-water-supply.net/_ressources/documents/default/1-861-2-1568125607.pdf> [accessed 20 September 2020].

Rushton, K.R., Eilers, V.H.M. and Carter, R.C. (2006) 'Improved soil moisture balance methodology for recharge estimation', *Journal of Hydrology* 318: 379–99 <https://doi.org/10.1016/j.jhydrol.2005.06.022>.

Ryan, P. (2008) 'Stepping into action: the second report on Citizens' Action for accountability in water and sanitation', WaterAid Report, July. <https://www.ircwash.org/sites/default/files/Ryan-2008-Stepping.pdf> [accessed 26 September 2020].

Sansom, K. and Koestler, L. (2009) 'African handpump market mapping study', Summary Report for UNICEF WASH Section and Supply Division, October. <https://wedc-knowledge.lboro.ac.uk/resources/pubs/Handpump_market_study_summary_report.pdf> [accessed 3 December 2020].

Schmidt, W-P. (2014) 'The elusive effect of water and sanitation on the global burden of disease', *Tropical Medicine and International Health* 19 (5): 522–27 <https://doi.org/10.1111/tmi.12286>.

Schouten, T. and Moriarty, P. (2003) *Community Water, Community Management; From System to Service in Rural Areas*, London: ITDG Publishing. <https://www.ircwash.org/sites/default/files/205.1-03CO-18010.pdf> [accessed 20 September 2020].

SDG Partnership Platform (no date, a) 'Accelerating progress towards SDG 6: a system strengthening approach for water, sanitation and hygiene that leaves no one behind' [online]. <https://sustainabledevelopment.un.org/partnership/?p=30208> [accessed 4 September 2020].

SDG Partnership Platform (no date, b) 'BabyWASH coalition' [online]. <https://sustainabledevelopment.un.org/partnership/?p=11856> [accessed 14 September 2020].

Sen, A. (1981) Poverty and Famines: An Essay on Entitlement and Deprivation, Oxford: Clarendon Press.

Sen, A. (1987) *Hunger and Entitlements*, Helsinki: World Institute for Development Economics Research. <https://www.wider.unu.edu/sites/default/files/RFA-1.pdf> [accessed 15 July 2020].

Sen, A. (1999) *Development as Freedom*, Oxford: Oxford University Press.

Shaw, R. (2015) *Drawing Water: A Resource Book of Illustrations on Water, Sanitation, Health, Hygiene, Rescue and Care in Low-Income Countries*, 2nd edn, Loughborough: Water Engineering and Development Centre. <https://wedc-knowledge.lboro.ac.uk/resources/e/dw/index.html> [accessed 17 September 2020].

Shullman, S.L. (2020) '"We are living in a racism pandemic," says APA president' [press release], *American Psychological Association*, 29 May. <https://www.apa.org/news/press/releases/2020/05/racism-pandemic> [accessed 15 July 2020].

Sorensen, J.P.R., Lapworth, D.J., Marchant, B.P., Nkhuwa, D.C.W., Pedley, S., Stuart, M.E., Bell, R.A., Chirwa, M., Kabika, J., Liemisa, M. and Chibesa, M. (2015) 'In-situ tryptophan-like fluorescence: a real-time indicator of faecal contamination in drinking water supplies', *Water Research* 81: 38–46 <https://doi.org/10.1016/j.watres.2015.05.035>.

Ssenyonjo, M. (2017) 'The influence of the international covenant on economic, social and cultural rights in Africa', *Netherlands International Law Review* 64: 259–89 <https://doi.org/10.1007/s40802-017-0091-4>.

Stelmach, R.D. and Clasen, T. (2015) 'Household water quantity and health: a systematic review', *International Journal of Environmental Research and Public Health* 12 (6): 5954–74. <https://doi.org/10.3390/ijerph120605954>.

Sustainable Development Goals (SDG) Tracker (no date) 'Sustainable development goal 6: ensure access to water and sanitation for all'. <https://sdg-tracker.org/water-and-sanitation> [accessed 7 July 2020].

Sutton, S. with Butterworth, J. (2021) *Self-Supply: Filling the Gaps in Public Water Supply Provision*, Rugby: Practical Action Publishing <https://practicalactionpublishing.com/book/2530/self-supply>.

Swain, B., Wicken, J. and Ryan, P. (2006) 'Citizens' action: how bridging the accountability gap leads to improved services', in J. Fischer (ed.) *Sustainable Development of Water Resources, Water Supply and Environmental Sanitation: Proceedings of the 32nd WEDC International Conference*, pp. 173–76, Loughborough: Loughborough University. <https://dspace.lboro.ac.uk/2134/28629> [accessed 26 September 2020].

Thomas, E., Andres, L.A., Borja-Vega, C. and Sturzenegger, G. (eds) (2018) Innovations in WASH Impact Measures: Water and Sanitation Measurement Technologies and Practices to Inform the Sustainable Development Goals, Washington, DC: World Bank. <https://openknowledge.worldbank.org/bitstream/handle/10986/29099/9781464811975.pdf?sequence=4&isAllowed=y> [accessed 23 July 2020].

Thompson, J., Porras, I., Katui-Katua, M., Mujwahuzi, M. and Tumwine, J. (2001) 'Drawers of water II: 30 years of change in domestic water use & environmental health in east Africa', IIED Report. <https://pubs.iied.org/pdfs/9049IIED.pdf> [accessed 15 July 2020].

Transparency International (2020) 'Together against corruption: Transparency International Strategy 2020'. <https://images.transparencycdn.org/images/TogetherAgainstCorruption_Strategy2020_EN.pdf> [accessed 17 July 2020].

Trevett, A.F., Carter, R.C. and Tyrrel, S.F. (2005a) 'The importance of domestic water quality management in the context of faecal-oral disease transmission', *Journal of Water and Health* 3 (3): 259–70 <https://doi.org/10.2166/wh.2005.037>.

Trevett, A.F., Carter, R.C. and Tyrrel, S.F. (2005b) 'Mechanisms leading to post-supply water quality deterioration in rural Honduran communities', *International Journal of Hygiene and Environmental Health* 208, 153–61 <https://doi.org/10.1016/j.ijheh.2005.01.024>.

Tucker J., MacDonald A.M., Coulter L. and Calow R.C. (2014) 'Household water use, poverty and seasonality: wealth effects, labour constraints, and minimal consumption in Ethiopia', *Water Resources and Rural Development* 3: 27–47 <https://doi.org/10.1016/j.wrr.2014.04.001>.

Tutusaus, M. and Schwartz, K. (2020) 'Commercialisation as organised hypocrisy: the divergence of talk and action in water services in small towns in Uganda', *Water Alternatives* 13 (2): 248–65. <http://www.water-alternatives.org/index.php/alldoc/articles/vol13/v13issue2/573-a13-2-3/file> [accessed 25 September 2020].

United Nations (UN) (no date, a) 'Democracy' [online]. <https://www.un.org/en/sections/issues-depth/democracy/> [accessed 15 July 2020].

UN (no date, b) 'Human rights to water and sanitation' [online]. <https://www.unwater.org/water-facts/human-rights/> [accessed 9 July 2020].

UN (1948) *Universal Declaration of Human Rights*. <https://www.un.org/en/udhrbook/pdf/udhr_booklet_en_web.pdf> [accessed 17 July 2020].

UN (1977) 'Mar del Plata Action Plan', United Nations Water Conference, Mar Del Plata, Argentina, 14–25 March. <https://www.internationalwaterlaw.org/bibliography/UN/UN_Mar%20del%20Plata%20Action%20Plan_1977.pdf> [accessed 16 September 2020].

UN (1990) 'New Delhi statement'. <http://www.ielrc.org/content/e9005.pdf> [accessed 3 December 2020].

UN (2000) 'United Nations Millennium Declaration: Resolution adopted by the General Assembly', A/RES/55/2. <https://undocs.org/en/A/RES/55/2> [accessed 20 September 2020].

UN (2002) 'General comment no. 15: the right to water', United Nations Committee on Economic, Social and Cultural Rights, E/C.12/2002/11. <https://digitallibrary.un.org/record/486454?ln=en> [accessed 14 July 2020].

UN (2019) 'Leaving no-one behind', World Water Development Report 2019, UNESCO, Paris. <https://en.unesco.org/themes/water-security/wwap/wwdr/2019> [accessed 25 September 2020].

United Nations Department of Economic and Social Affairs (UN DESA) (2018) 'World urbanization prospects 2018', Population Division, New York, NY. <https://population.un.org/wup/> [accessed 14 September 2020].

UN DESA (2019a) 'World economic situation and prospects, statistical annex', Population Division, New York, NY. <https://www.un.org/development/desa/dpad/wp-content/uploads/sites/45/WESP2019_BOOK-web.pdf> [accessed 14 July 2020].

UN DESA (2019b) 'World population prospects', Population Division, New York, NY. <https://population.un.org/wpp/> [accessed 14 July 2020].

UN DESA (2019c) 'World urbanization prospects: the 2018 revision', Population Division, New York, NY. <https://population.un.org/wup/Publications/Files/WUP2018-Report.pdf> [accessed 14 July 2020].

United Nations Environment Programme (UNEP) (2010) Africa Water Atlas, Malta: Progress Press Ltd. <http://wedocs.unep.org/bitstream/handle/20.500.11822/7919/africa_water_atlas.pdf?sequence=3&isAllowed=y> [accessed 12 September 2020].

United Nations International Children's Emergency Fund (UNICEF) (2010) Professionalizing the Manual Drilling Sector in Africa: A Guide to Building Capacity to Increase Access to Safe Water in Rural Areas, New York, NY: UNICEF/Enterprise Works/VITA/Practica Foundation. <https://www.unicef.org/wash/files/1._Proffessionalizing_Manual_Drilling_lo_res.pdf> [accessed 12 September 2020].

UNICEF (2016) 'Strengthening the enabling environment for water, sanitation and hygiene (WASH)', Guidance note, May. <https://www.unicef.org/wash/files/WASH_guidance_note_draft_10_3_hr.pdf> [accessed 9 July 2020].

UNICEF (2017) 'Sustainability checks: guidance to design and implement sustainability monitoring in WASH'. <https://www.unicef.org/wash/files/WASH_Guidance-for-Sustainability-Checks_.pdf> [accessed 16 July 2020].

UNICEF (2018) 'Global evaluation of UNICEF's drinking water supply programming in rural areas and small towns, 2006–2016', Evaluation report, October. <https://www.unicef.org/evaldatabase/index_103361.html> [accessed 23 July 2020].

UNICEF (2020) 'UNICEF WASH programme contribution to COVID-19 prevention and response'. <https://www.unicef.org/sites/default/files/2020-03/UNICEF_WASH_Programme_COVID-19_prevention_response.pdf> [accessed 14 July 2020].

UNICEF and SIWI (2019) 'WASH bottleneck analysis tool: country implementation guide', <https://www.washbat.org/wp-content/uploads/pt/WASH_BAT_Implementaton_Guide_Website.pdf> [accessed 16 July 2020].

United Nations General Assembly (2015) 'Transforming our world: the 2030 Agenda for Sustainable Development', Resolution adopted by the General Assembly on 25 September 2015, A/RES/70/1. <https://www.un.org/ga/search/view_doc.asp?symbol=A/RES/70/1&Lang=E> [accessed 7 July 2020].

UN-Water (no date) 'Human rights to water and sanitation' [online]. <https://www.unwater.org/water-facts/human-rights/> [accessed 7 July 2020].

UN-Water (2019) 'National systems to support drinking-water, sanitation and hygiene: global status report 2019', UN-Water Global Analysis and Assessment of Sanitation and Drinking-Water (GLAAS) 2019 Report, Geneva: WHO. <https://www.who.int/water_sanitation_health/publications/glaas-report-2019/en/> [accessed 25 September 2020].

UN-Water and WHO (2015) 'UN-Water GLAAS TrackFin Initiative: tracking financing to sanitation, hygiene and drinking-water at the national level'. <https://apps.who.int/iris/bitstream/handle/10665/204861/WHO_FWC_WSH_15.23_eng.pdf?sequence=1> [accessed 16 July 2020].

USAID (2018) Toward a Hygienic Environment for Infants and Young Children: A Review of the Literature. Washington, DC: USAID Water, Sanitation, and Hygiene Partnerships and Learning for Sustainability (WASHPaLS) Project. <https://pdf.usaid.gov/pdf_docs/PA00SV56.pdf> [accessed 14 August 2020].

Valcourt, N., Javernick-Will, A., Walters, J. and Linden, K. (2020) 'System approaches to water, sanitation, and hygiene: a systematic literature review', *International Journal of Environmental Research and Public Health* 17 (3): 702 <https://doi.org/10.3390/ijerph17030702>.

Valcourt, N., Walters, J., Javernick-Will, A., Linden, K. and Hailegiorgis, B. (2020) 'Understanding rural water services as a complex system: an assessment of key factors as potential leverage points for improved service sustainability', *Sustainability* 12: 1243 <https://doi.org/10.3390/su12031243>.

van de Giessen, E. (no date) 'Solar pumping for village water supply systems: training manual', Practica Foundation, The Netherlands. <https://www.practica.org/wp-content/uploads/2017/05/Solar-Pumping-design-manual-eng-1.pdf> [accessed 15 September 2020].

van der Kerk, A. (2015) 'Quick guide to costs', IRC WASH. <https://www.ircwash.org/news/costs> [accessed 23 September 2020].

van Genuchten, M.Th. and Pachepsky, V. (2011) 'Hydraulic properties of unsaturated soils', in J. Glinski, J. Horabik and J. Lipiec (eds), *Encyclopaedia of Agrophysics*, Dordrecht: Springer <https://doi.org/10.1007/978-90-481-3585-1_69>.

VanDerslice, J. and Briscoe, J. (1995) 'Environmental interventions in developing countries: interactions and their implications', *American Journal of Epidemiology* 141 (2): 135–44 <https://doi.org/10.1093/oxfordjournals.aje.a117401>.

Velis, M., Conti, K.I. and Biermann, F. (2017) 'Groundwater and human development: synergies and trade-offs within the context of the sustainable development goals', *Sustainability Science* 12: 1007–17 <https://doi.org/10.1007/s11625-017-0490-9>.

Visscher, J.T. (2006) 'Facilitating community water supply treatment: from transferring filtration technology to multi-stakeholder learning', PhD thesis, Wageningen Universiteit. <https://edepot.wur.nl/121780> [accessed 11 September 2020].

Voltaire (1764) *Voltaire's Philosophical Dictionary*. <https://www.gutenberg.org/files/18569/18569-h/18569-h.htm> [accessed 9 July 2020].

Wagner, E. and Lanoix, J. (1958) *Excreta Disposal for Rural Areas and Small Communities*, Monograph Series No. 39, Geneva: WHO.

Wagner, E.G. and Lanoix, J.N. (1959) *Water Supply for Rural Areas and Small Communities*, Monograph 42, Geneva: WHO. <https://www.who.int/water_sanitation_health/dwq/monograph42.pdf> [accessed 15 September 2020].

Walters, J.P. and Javernick-Will, A.N. (2015) 'Long-term functionality of rural water services in developing countries: a system dynamics approach to understanding the dynamic interaction of factors', *Environmental Science and Technology* 49: 5035–43 <https://doi.org/10.1021/es505975h>.

Warner, D.B., Briscoe, J., Hafner, C. and Zeilmer, B. (1986) 'Malawi self-help rural water supply program: final evaluation', WASH field report no. 186. <https://www.ircwash.org/sites/default/files/824-6995.pdf> [accessed 15 September 2020].

WASHCost (2012) 'Providing a basic level of water and sanitation services that last: cost benchmarks', WASHCost Infosheet 1, October. <https://www.ircwash.org/sites/default/files/IRC-2012-Providing.pdf> [accessed 25 September 2020].

Water Engineering and Development Centre (WEDC) (no date) 'The Blair bucket pump', Poster 34. <https://wedc-knowledge.lboro.ac.uk/resources/posters/P034_The_Blair_bucket_pump.pdf> [accessed 17 September 2020].

Water Integrity Network (no date) 'What is water integrity?' [online]. <https://www.waterintegritynetwork.net/water-integrity-basics/> [accessed 17 July 2020].

Water Point Data Exchange (WPDx) (no date) 'The water point data exchange' [online]. <https://www.waterpointdata.org> [accessed 17 September 2020].

Water and Sanitation for Health Project (1993) *Lessons Learned in Water, Sanitation and Health: Thirteen Years of Experience in Developing Countries*, Washington, DC: USAID. <http://www.ehproject.org/PDF/ehkm/wash-lessons_learned1993.pdf> [accessed 20 September 2020].

Water and Sanitation Programme (WSP) (2002) 'Upgraded family wells in Zimbabwe: household-level water supplies for multiple uses', Field note 6, World Bank, Nairobi. <https://waterfund.go.ke/watersource/Downloads/001.%20Upgraded%20Family%20Wells%20in%20Zimbabwe.pdf> [accessed 16 September 2020].

WaterAid (no date) 'SusWASH' [website]. <https://washmatters.wateraid.org/suswash> [accessed 29 September 2020].

WaterAid (2011) 'Sustainability framework' [online]. <https://washmatters.wateraid.org/sites/g/files/jkxoof256/files/sustainability%20framework.pdf> [accessed 23 September 2020].

WaterAid (2019) 'WaterAid glossary of key terms – system strengthening and empowerment' [online], November. <https://washmatters.wateraid.org/sites/g/files/jkxoof256/files/wateraids-system-strengthening-and-empowerment-glossary.pdf> [accessed 9 July 2020].

WaterAid (2020) 'From data to decisions: developing user-centred monitoring programmes for water, sanitation and hygiene', Policy brief, June. <https://washmatters.wateraid.org/sites/g/files/jkxoof256/files/from-data-to-decisions-developing-user-centred-monitoring-programmes-for-water-sanitation-and-hygiene-policy-brief.pdf> [accessed 11 November 2020].

WEDC/WaterAid (no date) 'Facilitator notes: accessibility and safety audit of water & sanitation facilities' [online]. <http://wedc.lboro.ac.uk/resources/learning/EI_FN2_Accessibility_Safety_Audit_v2.pdf> [accessed 25 September].

Wenar, L. (2006). 'Accountability in international development aid', *Ethics & International Affairs* 20 (1): 1–23 <https://doi.org/10.1111/j.1747-7093.2006.00001.x>.

Whaley, L. and Cleaver, F. (2017) 'Can "functionality" save the community management model of rural water supply?' *Water Resources and Rural Development* 9: 56–66 <http://dx.doi.org/10.1016/j.wrr.2017.04.001>.

Whaley, L., MacAllister, D.J., Bonsor, H., Mwathunga, E., Banda, S., Katusiime, F., Tadesse, Y., Cleaver, F. and MacDonald, A. (2019) 'Evidence, ideology and the policy of community management in Africa', *Environmental Research Letters* 14 (8) <https://doi.org/10.1088/1748-9326/ab35be>.

Wilby, R.L. and Dessai, S. (2009) 'Robust adaptation to climate change', *Weather* 65 (7): 180–5 <https://doi.org/10.1002/wea.543>.

Wisner, B. and Adams, J. (eds) (2002) *Environmental Health in Emergencies and Disasters: A Practical Guide*, Geneva: WHO. <https://apps.who.int/iris/handle/10665/42561> [accessed 26 September 2020].

World Bank (2015) *World Development Report: Mind, Society and Behavior*, Washington, DC: World Bank Group. <https://www.worldbank.org/en/publication/wdr2015> [accessed 26 September 2020].

World Bank (2017a) 'Sustainability assessment of rural water service delivery models: findings of a multi-country review', Washington, DC. <https://openknowledge.worldbank.org/bitstream/handle/10986/27988/W17055.pdf?sequence=4&isAllowed=y> [accessed 16 September 2020].

World Bank (2017b) 'Toward a universal measure of what works on rural water supply: rural water metrics global framework', Water Global Practice, WSP topic brief, World Bank Group. <https://www.issuelab.org/resources/30579/30579.pdf?download=true> [accessed 24 July 2020].

World Bank (2018) *Solar Pumping: The Basics*. Washington, DC: World Bank. <http://documents.worldbank.org/curated/en/880931517231654485/pdf/123018-WP-P159391-PUBLIC.pdf> [accessed 15 September 2020].

World Bank (2020a) 'Global economic prospects' [online], June <https://doi.org/10.1596/978-1-4648-1553-9>.

World Bank (2020b) 'Poverty and equity databank' [online]. <https://databank.worldbank.org/reports.aspx?source=poverty-and-equity-database#> [accessed 7 June 2020].

World Bank (2020c) 'World Bank country and lending groups' [online]. <https://datahelpdesk.worldbank.org/knowledgebase/articles/906519-world-bank-country-and-lending-groups> [accessed 15 May 2020].

World Bank (2020d) 'Databank' [online]. <https://databank.worldbank.org/home.aspx> [accessed August 2020].

World Health Organization (WHO) (no date) 'Water-related diseases' [online], <https://www.who.int/water_sanitation_health/diseases-risks/diseases/arsenicosis/en/> [accessed 11 September 2020].

WHO (1984) *Guidelines for Drinking Water Quality*, Vol. 1 Recommendations, Geneva: WHO.

WHO (2012) *Accelerating Work to Overcome the Global Impact of Neglected Tropical Diseases: A Roadmap for Implementation*, Geneva: WHO. <https://unitingtocombatntds.org/wp-content/uploads/2017/11/who_ntd_roadmap.pdf> [accessed 10 September 2020].

WHO (2014) 'Antimicrobial resistance: an emerging water, sanitation and hygiene issue', Briefing note, WHO/FWC/WSH/14.7 <https://apps.who.int/iris/bitstream/handle/10665/204948/WHO_FWC_WSH_14.7_eng.pdf?sequence=1> [accessed 11 September 2020].

WHO (2015) 'Water sanitation and hygiene for accelerating and sustaining progress on neglected tropical diseases: a global strategy, 2015–2020'. <https://apps.who.int/iris/bitstream/handle/10665/182735/WHO_FWC_WSH_15.12_eng.pdf?sequence=1> [accessed 10 September 2020].

WHO (2016) *Results of Round I of the WHO International Scheme to Evaluate Household Water Treatment Technologies*, Geneva: WHO. <https://apps.who.int/iris/bitstream/handle/10665/204284/9789241509947_eng.pdf?sequence=1> [accessed 10 September 2020].

WHO (2017a) *Guidelines For Drinking Water Quality*, 4th edn, Geneva: WHO. <https://www.who.int/water_sanitation_health/publications/drinking-water-quality-guidelines-4-including-1st-addendum/en/> [accessed 12 September 2020].

WHO (2017b) *Safely Managed Drinking Water*, Geneva: WHO and UNICEF. <https://washdata.org/report/jmp-2017-tr-smdw> [accessed 20 September 2020].

WHO (2018) *WASH and Health Working Together: A 'How-To' Guide for Neglected Tropical Disease Programmes*, Geneva: WHO. <https://apps.who.int/iris/bitstream/handle/10665/279913/9789241515009-eng.pdf?ua=1> [accessed 11 September 2020].

WHO (2020a) *Ending the Neglect to Attain the Sustainable Development Goals: A Road Map for Neglected Tropical Diseases 2021–2030*, Geneva: WHO. <https://www.who.int/neglected_diseases/Revised-Draft-NTD-Roadmap-23Apr2020.pdf> [accessed 14 September 2020].

WHO (2020b) 'Trachoma' [online], <https://www.who.int/news-room/factsheets/detail/trachoma> [accessed 10 September 2020].

Young, K.G. (2019) 'Introduction', in K. Young (ed.), *The Future of Economic and Social Rights*, pp. 1–34, Cambridge: Cambridge University Press <https://doi.org/10.1017/9781108284653>.

Zhou, Y. (2009) 'A critical review of groundwater budget myth, safe yield and sustainability', *Journal of Hydrology* 370: 207–13 <https://doi.org/10.1016/j.jhydrol.2009.03.009>.

Index

Page numbers in **bold** refer to tables and in *italics* to figures.

accountability *see* WASH
Accra Agenda for Action 110
affordability 122–9
   community management and 116
   drinking water 7
   human right 162
   as indicator **178**, **179**
   national budgets and 128
   post-construction costs 181
Africa
   groundwater 30, 44
   population growth 45
   rural population 2
   Sahel paradox 33
   see also sub-Saharan Africa
Afridev pump 73, *74*, 80
Agenda for Change 190
agriculture 10, 23, 29, 37, 45, 136, 138, 186
animal excreta 17, *18*
antimicrobial resistance 26
Aquastat 30
aquifers 9, 31–2
   confined 38–40
   drawdown *see* boreholes
   fluoride and 22
   properties and boreholes 56
   storativity 38–41
   transmissivity 63–4
   unconfined *31–2*, 38–41, *43*
   water movement in 41–3
   yield 44
arsenic 21–3
asset management 125
associations (professional and technical) 49, *158*, 160, 164
at least basic service **5**, 167, 169–70, 172–3, 192

BabyWASH Coalition 17
bacteria 15, 17–18, 26
Bangladesh 16, 19, 21–2, 72, **202**, **204**
baseflow 36
basic water service 1, 4, 7, 10, 14–16, 25–8, 87, 121, 126, 128, 132, 138–9, 148, 167–173, 179, 191–3
benefit/cost ratio 121, 131
Bill and Melinda Gates Foundation 122, 181
biofouling 58
Blair Research Institute 70, 75
Black Lives Matter xxi
blind (visually impaired) 15
Blue Pump 75–7
blue water 37–8
Boode (company) 77
boreholes 49–66
   aquifers and 39–40
   casing 53–60, 66, 91
   completion 59–62
   cost-effectiveness of 51, 65–6
   depth 49–50, 53, 57, 59–60, 65–6, 145
   design 53–8, 66
   diameter 50, 53, 55–8, 67, 91
   discharge 53, 56–8, 63–4
   drawdown 40, 42–4, 53, 55–6, 58, 63, 94–5
   drilling 58–9
   efficiency 63
   pumping tests 41–2, *43*
   site selection 51–3
   specifications 62–5
   well development 53, 62
   well screens 53–4, 57–9, 62, 64, 91
   yield 44, **45**, 47

bottleneck analysis 151, 164
Bucket Pump 70, *71*

Canzee pump 71, *72*
capex (capital expenditure) 2, 6, 8, 96–8, 123–129, 132, 181
capital cost *see* capex
capital investment *see* capex
capmanex (capital maintenance expenditure) 96–7, **124**–7, 129, 132–3
Central and South America 70, 72
Central and Southern Asia *14*, 26
children
    disease 1–2, 15, 17–19
    inequality 140
    poverty 138
    water carrying 4, 24, 131
Chlamydia trachomatis 15
chlorination 23, 97
cholera 18, 108
Citizens' Action 161
climate change 9, 32, 35, 44–5, 47, 123
colonialism 199
community-based maintenance *see* community management
community engagement 146–7
community management 2–3, 114–20, 196
    demographics and 45
    finance 145
    functionality 115–16, 119
    monitoring 144
    piped systems 143
    repairs and 7
    solar water pumping and 97
    weaknesses 9, 114, 146, 180
community management plus 115
community participation 2, 79, 117, 144–6, 183
community water supply 6, 135–50, 160–4
    demographic trends and 142–3
    equity and inclusion 147–50
    effectiveness 143–5
    organization 137–8
    participation 145–6
    payment and usage 155–6
    poverty and disadvantage 138–41

Community Water Supply: The Handpump Option (UNDP and the World Bank) 79
complex adaptive systems 183–4
complexity 195, 198
compliance and coordination 194
cone of depression *see* boreholes: drawdown
conservation of mass 41
constant discharge test 63
consumption 7, 15–17, 36, 38, 45, 108
consumptive and non-consumptive uses of water 38
corrosion 58, 81–2, *83*, 88, **175**
corruption 111, 156–7, 166
cost-effectiveness 50, 51, 65–6, 117, 164
countries, categorization of 202–5
country status overviews 164
Covid-19 xix, xxi

Darcy's Law 41
decentralization of governance 110, 114, 194
Degree of Urbanization 201–2
defecation 16–17, 191, 198
demand 100, 144–5
democratic governance 162–3
demographic transition 142
demographic trends 142–3
Design Outreach (non-profit organization) 77
DFID (Department for International Development), UK 165
diarrhoea 1–2, 15–19, 108, 131
direct action pumps 71–3
disabled people 25, 150
discrimination 140–1, 147–9, *159*
disease 1–2, 15, 17–19
down-the-hole hammer *61*
down time 84–5
drawdown 40, 42–4, 56, 58, 63, *94*
drillers' associations 49, 59
drilling 51–3, 56–63, 65–6
    contracts and specifications 50–2, 54, 62, 65, 82, 182
    manual 49, 58–60, 107
    mechanical 58–9, 107
    procurement 12, 51, 123

drinking water
  benefit/cost ratio *131*
  chemistry 20–3, 81, 97
  community management 2–3
  coverage **5**
  finance 128, 131–2
  groundwater 36, 37
  as human right 162
  lack of 1
  multi-dimensional poverty index (MDPI) 138–9
  quality and health 17–20
  recontamination 24
  requirements 15–17, 109, 110
  sampling 64
  schools and health facilities 25–6
  service levels 3–4, 7–8, **168**
  solar-powered 99–101
  surface water 27–8
  Sustainable Development Goal (SDG) 152, 190
  treatment 23–4, 107–8
  water carrying 24–25

education 19, 26, 138–9, 159
EED (environmental enteric dysfunction) 19
effectiveness 23, 80, 143–5, 154, 157
elderly people 25, 150
electric submersible pumps 90–4
electricity 69, 86, 90, 93–4, 139
EMAS direct action pump 71–2
empathy 188, 198
encrustation 58, 63, 88
energy 4, 16, 24, 58–9, 68–9, 77, 86, 88, 90, 92, 94–5, 98, 123, 136
England and Wales 128, 155
entrance velocity 58
environmental enteric dysfunction *see* EED
EPA-Net 88
equality and inclusion 147–50
Escherichia coli 18
ethics, code of 189–90
ethos 117, 153, 159
Ethiopia 16, 18–19, 84–5, 137, 171, *172*, **202**, **204**
European Union 123
evaporation 22, 31–2, 35

evapotranspiration *32*, 35–6
excreta 13, 17–18, 26
exclusion xxi, 20, 135, 137, 140, 147
externally initiated supply 109–12
eye disease 15

facial cleanliness 15
faecal
  coliforms 18
  contamination 17–19, 23, 28, 62, 108
  indicator bacteria 18
faecal-oral (transmission route) 13, 17, 19
Fairwater Foundation 77
FAO (Food and Agriculture Organization of the United Nations) 30
F-diagram 17–18
finance 121–34
  affordability 122–9
  life-cycle costing (LCC) 122–3, **124**
  national budget allocations 194
  trends 181–3
  viability 132–3
first 1,000 days 17, 19
fluoride 21, 22
fluorosis 22
Food and Agriculture Organization of the United Nations *see* FAO
functionality 8, 20, 28, 51, 67, 82, 84–5, 115–16, 119–20, **124**, 133, 141, 177, 179

galvanized components 71, 81–3, 144
gender-based violence 25, 145
gender equality 140–1, 145, **149**
geogenic contamination 13, 21
geophagy 17–18
geophysics 52, **175**
Ghana 44, 72, 122, **170**, 171, **203**, **204**
GLAAS (UN-Water Global Analysis and Assessment of Sanitation and Drinking-water) 110, 128, 176–7
GNI (gross national income) *173*, 174, 202–3

governance 112, 137, 162–6, 176, **179**, 180, 194
governments 156, 158–9
  capital investments 6
  cost-effective boreholes 51
  equality and inclusion 150
  finance 125, 127–8, 132, 134, 154, 180–1, 186–8
  local 84, 85, 90, 107, 144, 177, 194
  management of water service 107–15
  mandate of 189–90
  monitoring 176–7
  standardization and 80
  support for 163–4
  system strengthening 197
gravity flow schemes 88–90, 117
green water 37–8
greenhouse gases 47, 123
grey water 38
gross national income see GNI
groundwater 29–47
  abstraction rates 44
  agricultural abstractions of 37
  chemistry of 21
  climate change 47
  estimation of 30
  flows and pumping 42, 43
  population 45–7
  recharge 31, 32–3, 36–8, 44, 47
  resources 9, 11, 29–33, 36–45, 47
  storage and movement 38–43
  unimproved 4, 6, 28
  water balance 31–3
  and water chemistry 21–3
  water cycle 30–1, 33–8

Handpump Technology Network 80
handpumps 67–86
  breakdown 118
  components 68
  corrosion 81–2, 83
  deep well 72–7
  functionality 82, 84–5
  'Hidden Crisis' project 84–5
  low-lift 70–2
  public/private domain 80
  quality of component parts 67, 73, 81–2, 86, 119

renewable energy sources 69
Sustainable Development Goal (SDG) 85–6
standardization 80–1
types 69–77
village-level operation and maintenance (VLOM) 79–80
hand washing 16–17, 155
health 13–28
  antimicrobial resistance 26
  and finance 130–2
  multi-dimensional poverty and 139
  schools 25–6
  water-carrying 24–5
  water quality 17–23
  water treatment 23–4
health care facilities 26, **27**
Hidden Crisis project 84–5, 144
high-income country 7, 110, 129, 173, 202
house connections 6, 87–8, 103, 131
household water treatment 23–4, 116
Household Water Treatment and Safe Storage Network see HWTS
human right, water as 7–8, 109, 147–8, 156, 158, 160–3, 186–8
Human Rights Council 147
HWTS (Household Water Treatment and Safe Storage) Network 24
hydraulic
  conductivity 40–2, 52, 56
  gradient 41
hydrogeological surveys 52
hydrological cycle see water cycle
hygiene
  hand washing with soap 155
  health care facilities **27**
  neglected tropical diseases (NTD) and 20
  permanent change in 123
  schools 25–6
  water quality and 18
  water quantity and 15–17, 108, 162

IBNET (International Benchmarking Network) 177, **178**

INDEX 243

ICESCR (International Covenant on Economic, Social and Cultural Rights) 187
imagination 12, 153, 183, 185–8, 198
Impact Pump **70**, 98, *99*, **175**
improved water points 2, 4, 10, 25, 27, 28, 138–9, 168–9, 191–2
improved water services 1, 4, 7, 12, 14, 16, 26, 121, 130–1, 138, 155, 168–9, 191
India 16, *37*, 72–3, 79, 122, **203**
India Mark II pump 73, *74*, 75, 80–1, *83*
India Mark III pump 73
inequality and exclusion 140–1
infrastructure 6, 87–103, **179**
  community management of 2–3, 143
  finance 121, 125, 132, 181
  gravity flow schemes 88–90
  Impact Pumps 98
  mechanically pumped systems 90–2, *93*
  piped distribution 101–3
  public standposts 103
  solar pumping 93–100
  sustainability of 189
  technology 174–6
  water storage 100–1
insecurity
  food 19, 138
  water 138
institutional reform *158*, 165–6
International Benchmarking Network *see* IBNET
International Covenant on Economic, Social and Cultural Rights *see* ICESCR
International Decade for Action on Water for Sustainable Development 188
International Water Association 177
IRC (WASH organization) 122–3, 164, 181, 197
irrigation 4, 28, 36–7, 44–5, 56, **168**
iron 22, 81–2

Jansen Venneboer B.V. 75
jetting 59, *60, 61*

JMP (Joint Monitoring Programme), UNICEF/WHO
  definitions 4, 10, 138–9
  service level 1, 5, 14, 16, 155, 168, **169-71**, 173, 176–7, 192
  service level ladder 25–6, **27**
joint sector reviews 164
justiciability 187

Kanchan water filter 23
KDWSP (Kigezi Diocese Water and Sanitation Programme) 117, 153–4
Kenya 19, 71, 73, 77, **170**, 171, **203**, **204**
kiosks 87–8, 94, 98, 101, 103

latrines **27**
LCC (life-cycle costing) 122–3, **124**
leadership 112, 117, 137–8, 157, *159*, 160, 189–90
levels of service 168–74
LifePump 77, **175**
limited water service 4–8, 12, 14, 16, 25–8, 138–9, 168–9, 191–3
Logan's formula 56
low-income country xix, 6, 26, 90, 107, 116, 127, 135, 138–42, 156, 158, 173, **178**, 184, 199, 202, 205

Madzi Alipo programme 118–19
Malawi
  classification **203**
  community water projects 146–7
  groundwater 45
  pumps 71, 73, 118–19
  service level **170**, 171, **172**, *173*
  water management 84–5
  water quality 19
Malawi Rural Water Supply Program 89–90
Maldev handpump 73
manganese 22–3
manual augering *60*
manual drilling 59
MDG (Millennium Development Goal) 2, 106, 130

MDPI (multi-dimensional poverty index) 138–40
mechanical drilling 59
menstrual hygiene 16, 25–6, **27**
methaemoglobinaemia 23
middle-income country 6, 15, 18, 64, 105, 113, 129, 135, 142–3, 153, 160, 166, 173, 184, 199, **202**, 205
Millennium Development Goal *see* MDG
monitoring
  climate change and 47
  cost-effectiveness 51
  equality and **149**
  investment 192, 194
  regulation and 164, 194
  technology and 8, 119, 174–9
  see also JMP (Joint Monitoring Programme)
Mono pump **70**, 77, *78*
mortality xix, 1, 13, 130–1, 142
mud rotary 59, *61*
multi-dimensional poverty index *see* MDPI

national income 173–4, 202
national water policies 193–4
negative individual attributes 159–60
neglected tropical diseases *see* NTD
Netherlands 75
Netherlands International Cooperation Collection 69
NGOs (non-governmental organizations) 6, 106, 108, 111–12, 132, 144, 148, 156, 158
Nicaragua 16, 70, **170**, 171, **202**
Nigeria *34*, **170**, 171, **202**
Nira AF-85 pump 72
nitrate concentrations 23
no-water-no-pay contracts 65
non-governmental organizations *see* NGOs
NTD (neglected tropical diseases) 19–20

Oceania 5–6, 26, 28, **205**
OD (open defecation) 191

official development assistance 8, 122, 132, 182
on-premises water supply 167–9, *172*, 191–3
opex (operation and minor maintenance expenditure) 96–7, **124**, 125, **126**, **127**, 132–3
organic compounds 23

Paris Declaration on Aid Effectiveness (Paris Principles) 110–11
pathogen xx, 2, 17–19, 24, 62, 198
PBR ('Payment by Results') 182
percussion 59, *60*
permeability *31*, 40, 62, 64
pH value 81
philanthropic finance 8, 181–2
piped systems 6, 88, 95, 103, 131
population 2, 33, 36, 44–7, 142–3, 169
  projections *3*, *142*
porosity 38–9
post-construction costs 90, 127
potentiometric surface 39, 41
poverty 111, 129, 136–41, 199
  multi-dimensional 138–40, 143
  water poverty xv, 8, 138
Practica Foundation 100–1
precipitation 30–1
private operators 6, 9, 112–14, 116, 119–20, 133, 155
professionalization 151, 164, 180
progressive realization 8, 187
protected springs and wells 2, **4**, 28, 88, *89*
protozoa 17
pump
  discharge 53, 56–8, 63–4, 73, 77, 92, 94–5
  head 69, 77, 92, 94–5, 98
  efficiency 92, 98
pumping water level 55–6, 94–5

quality assurance/control *see* handpumps: quality of component parts

rainfall 30–8, 45, 47
  variability *34*

rainwater harvesting 87, 107–8, 116
recharge *see* groundwater recharge
reciprocal agreements 144
recontamination 24
recurrent costs 87, 106, 113, 121–2, 124–6, 132–3, 181, 185, 192–4
reform initiatives 165–6
regulation 11, 152, 164
renewable freshwater resources 30
results-based financing 181–2
rest water level 54–6, 63, *94*
Richardson and Cruddas (company) 73
roof-water harvesting *see* rainwater harvesting
Rope Pump 70–1
runoff *32*, 36–7, 47
rurality, definitions of 136
RWSN (Rural Water Supply Network) 52, 59, 65, 69, 80, 82, 113, 164

safely managed water xv, 13–14, 108, 122, 126–8, 130, 168
Sahel paradox 33
sanitary seals 24, 53, *54*, 59, 62
sanitary surveys 19
sanitation
    costs 122–3, 128, 130, 188
    economic benefits of 132
    facilities 25, 27, 139
    human right to 147, 161–2, 187
    link to disease 1–2
    management 146
    monitoring 176–7
    services 20, 109, 161–2
    see also hygiene
schistosomiasis 20, 28
schools 16, 24, 25–6
scorecards 155
SDG (Sustainable Development Goal) 3–4, 8, 10, 152
    community participation 145
    costs 125, 128, 182
    geographical regions **204-5**
    handpumps and 85–6
    'safely managed' supply 108
    service levels 16

systems strengthening 197
    timetable 14, 20, 132
self-supply 6, 107–9, 113, 116, 132
sensors 119, **175**, 177
service levels **4**, **25**, 26–8, 168–74
SHINE trial 19
Sierra Leone *34*, **170**, 171, **203**, **204**
slavery xxi, 199
sludging 59, *61*
soil water *32*, 35, *42*
solar water pumping 93–101, **175**
    complexity and management 97–8
    cost 95–7
    Impact Pump 98
    kiosks 98–100, 103
    water storage 100–1
specific capacity 55–6, 64
standposts, public *see* water points
storage coefficient *see* storativity
storativity 38–9, 44
stunting 2, 17, 19
sub-Saharan Africa
    finance 131–2, 182
    handpumps 70, 80–2, 86
    health facilities 26
    lack of water service 5–6, 14, 26
    population 2, *46*, 142
    Sustainable Development Goal (SDG) **204**
subsidies 8, 109, 134
surface water 191–3
    coverage **5**, 6, 26, 28
    flow 30, 36
    neglected tropical diseases (NTD) 20
    piped supplies 127
    quality 14
    re-use 38
    self-supply 116
    service levels 4, **168**, 169
    water cycle 31
sustainability
    of community management 115, 117, 144, 146
    financial 7–9, 97, 132, **179**, 181, 193–4
    and functionality 85

of solar water pumping 97
of water services 2, 6, 8, 11, 20, 98, 164, 177, 183, 189, 192
sustainability checks 151, 164
Sustainability Framework (WaterAid) 123
Sustainable Development Goal *see* SDG
'Sustainable Services at Scale' *see* Triple-S
systems 183–4, 195–6
systems strengthening 12, 111, 118, 152, 154, 166, 181, 190, 197, 199

Tara pump 72
tariffs (water charges)
  England and Wales 128–9
  high-income countries 7
  indicators **179**
  private operators 114, 119, 133
  public water points 103
  state and 122, 132, 134
  trends 8–9
taxation 8, 110, 122, 132, 133
technology 8–9, 174–6
test pumping 63–4
Thermofluidics Ltd (company) 98
thyroid effects 23
time saving 130–2
total dynamic head 94–6
trachoma 15, 20
transfers 110, 122, 132
transmissivity 42–4, 55–6, 63–4
transparency 117, 163, 165
Transparency International 166
transpiration 32, 35
Triple-S ('Sustainable Services at Scale') 181
tubewells *see* boreholes

Uganda
  classification **203**
  community management 144
  'Hidden Crisis' project 84–5
  pumps 71, 81–4
  service level **170**, 171, 192–3
  water quality 19

see also KDWSP (Kigezi Diocese Water and Sanitation Programme); UPGro
UN (United Nations)
  categorization of countries 203
  Department of Economic and Social Affairs (DESA) 201
  Food and Agriculture Organization (FAO) 30
  human rights 160–3
  International Drinking Water and Sanitation Decade 3, 79, 106, 109
  'Mar del Plata Action Plan' 109
  Statistical Commission 201
  Universal Declaration of Human Rights 161–2
  see also GLAAS (UN-Water Global Analysis and Assessment of Sanitation and Drinking-water)
UNDP/World Bank Handpump programme 79
UNICEF (United Nations International Children's Emergency Fund) 1, 59, 73, 81, 154, 164, 174, **175**, 197
unimproved water points 6, 8, 14, 25, 28, **138**, 169, 191–3
unimproved water services 4–6, 25, 27, **138**, 168–9, 193
unsaturated zone 38
UPGro (Unlocking the potential of Groundwater for the Poor) 84, 116, 144
Uptime initiative 119–20
urbanization 2, 11, 141–2, 201
utility
  indicators **178**
  management 119
utilitization 180

values, ethical 12, 117, 153–4, *159*, 163, 186–190, 198
Vergnet Hydro pump 77, *78*
violence 25, 199
viruses xix, 17

VLOM (village-level operation and maintenance) 73, 79–80
Voices of the Poor (World Bank) 158
Volanta pump 75–7

Wales, England and 128, 155
WASH (water, sanitation, and hygiene)
  accountability 111, 156, 157, *158*, 160, 163–6, 180, 190
  Agenda for Change 190
  equality and inclusion 150, 199
  financing of 110, 130, 181, 182
  health facilities 26, **27**
  human rights 148, 187
  monitoring 176
  national level 164
  neglected tropical diseases (NTD) 20
  problems 196
  schools 25–6
  stunting and lack of 19
  sustainability 123
  system strengthening 152, 153, 184, 197
WASH Benefits trial 19
WASHCost 122–7, 181
Water and Sanitation for Health Project 114–15
water balance 31–3, 44
water carrying 4, 24–5, 28, 186
water cycle 30–8, 44
Water Integrity Network 166
Water Mission 96, 97
water points 50, *89*, 103
  contamination 108
  distance to 10, 16–17, 28, 192
  functionality 82, 84–5, 115–16, 118, 144–5
  management 119–20, 138, 180
  piped distribution 101
  siting of 40, 49, 51–3, 65–6, 82, 145, 176
  technology 8, **175**, 177
  see also improved water points; unimproved water points
water quality
  boreholes 49, 52
  failures 14

gravity flow systems 88
health and 17–23
human right 162
Millennium Development Goal (MDG) target 2
recontamination 24
sampling 62, 64
self-supply 107–8
technology **175**
unimproved groundwater 28
water carrying 24–5, 28
water chemistry 20–3, 81
water treatment 23–4
WHO standard **84**
water quantity 15–17, 44–5, 109, 145, 162
water safety plans 19
water sources *see* water points
water table 31, 38–9, 41, *60*
water treatment 14, 20, 23–4, 38, 99, 107, 116
WaterAid 113, 152
  Citizens' Action Report 161
  'Equity and inclusion framework' 147
  Sustainability Framework 123
  SusWASH programme 197
well screens 53–4, *55*, 57–9, 62, 64
wells
  abstraction rates 44, **45**
  cost 126–7, 129
  design 53–4, 57–8
  development 63–5
  quality 28
  self-supply 107–8
  sites 40, 51–2
  water flows 30, 42
  see also boreholes; handpumps
WHO (World Health Organization) 1, 15, 18–22, 24, 84, 108, 164
wicked problems 186, 195–6
willingness to pay 12, 114, 151
women
  benefits of good supply 131, 145
  disadvantage 140–1, 143
  equality **149**
  water carrying 4, 24–5
  well site selection 52

women-headed households 136
see also gender-based violence; gender equality
World Bank
  on costs 125
  country classification 139, 173–4, 202–3
  management models 113
  service level indicators 177, 179
  solar water pumping 95, 96–8, 101
  UNDP/World Bank Handpump programme 79
  utility management 119
  see also *Voices of the Poor*; *World Development Report*
World Bank Atlas 202
World Development Report (World Bank) 159
World Health Organization *see* WHO

yield
  aquifer 44, 97
  borehole 40, 44–5, 52, 57, 62, 97

Zimbabwe 19, 70–1, 75, **170**, 171, **203**, **204**
Zimbabwe Bush Pump 73, 75